COSMOLOGY + 1

Readings from
SCIENTIFIC AMERICAN

COSMOLOGY + 1

With Introduction by
Owen Gingerich

Harvard University
and
Smithsonian Astrophysical Observatory

W. H. Freeman and Company
San Francisco

COVER ILLUSTRATION: A synthetic optical photograph of our Galaxy as it might appear to an observer in a distant external galaxy. Produced by S. Christian Simonson III of the University of Maryland. Permission for reproduction granted by A. Blaauw, Chairman, Board of Directors, *Astronomy & Astrophysics.*

Library of Congress Cataloging in Publication Data
Main entry under title:

Cosmology + 1.

Includes bibliographies and index.
1. Cosmology—Addresses, essays, lectures. 2. Stars—Evolution—Addresses, essays, lectures. I. Gingerich, Owen. II. Scientific American.
QB981.C823 523.1′08 77-1448
ISBN 0-7167-0043-3
ISBN 0-7167-0042-5 pbk.

Printed in the United States of America

9 8 7 6 5 4 3 2

CONTENTS

Note on cross-references to SCIENTIFIC AMERICAN *articles:* Articles included in this book are referred to by title and page number; articles not included in this book but available as Offprints are referred to by title and offprint number; articles not included in this book and not available as Offprints are referred to by title and date of publication.

PREFACE

Infinity, curved space, the big bang, red shifts of galaxies—these are the makings of modern cosmology. Armed with sophisticated mathematics and all-too-sparse observations, cosmologists have labored since the 1920s to construct a picture of the universe on its grandest scale. But the unfettered imaginations of yesterday's thinkers scarcely anticipated the surprises that modern observational science has brought forth: cataclysmic galactic nuclei, quasars, gigantic radio galaxies. Enormous in size and energy compared to our terrestrial station in space, yet only small units in the cosmos at large, these entities offer fresh clues to the violent history of the expanding universe. Perhaps someday they will help solve the mystery of how the explosion of space and energy began, and whether it will ever take place again.

This collection of articles from *Scientific American* describes various aspects of our search for an understanding of the universe as a whole. It includes descriptions of galaxies, of the pervasive background radiation, and of non-Euclidian space. It also includes two recent articles on black holes, those astonishing space-warps in miniature that may ultimately yield crucial hints about the curvature of space as a whole.

Is the universe a one-time happening? Or does it pulse in an unending cycle? This is today the greatest question of cosmology, but to call it the greatest question of all is to confuse size with significance. There are indeed many other important inquiries that challenge the imagination, and within the wide scope of astronomy there is one rival in particular: Are we alone? Are we the only intelligent life in this vast celestial frame? For this reason this collection is called *Cosmology + 1:* the extra article concerns the fascinating but so far unsuccessful search for life elsewhere. Like the question whether our universe has exploded in a one-and-only incarnation, the question whether intelligent beings inhabit the satellites of some distant stars is quite undecided and has called forth wild speculation. These questions are intellectual mainsprings that drive much astronomical research today. I hope the reader will find the search for their answers exciting, even though they may never be found.

January 1977 *Owen Gingerich*

COSMOLOGY + 1

COSMOLOGY + 1

INTRODUCTION

With the publication of Newton's *Principia* in 1687, the stage was set for a new kind of inquiry into the nature of the universe on its largest scale. Given the concept of Universal Gravitation, the obvious question was Why didn't all the stars draw themselves together into one grand fiery ball? The solution to this puzzle proved so elusive that cosmology simply went into hibernation for over two centuries.

Today, the idea of gravitational attraction leading to gravitational collapse plays a key role in our understanding of many astronomical phenomena. The sun was once an extended gaseous sphere as large as the present solar system. Warming as it shrank, it finally achieved sufficiently high internal temperatures to trigger nuclear reactions, which have temporarily balanced the powerful gravitational pull. Eventually, when these fuels have been exhausted, gravity will crush the sun's material into a smaller, denser ball.

Elsewhere, gravitational forces have drawn gaseous spheres that were more massive (and hence have evolved more rapidly) into dense neutron stars. Some of these betray their existence by projecting radio beams that sweep across the earth as the neutron stars spin. Called pulsars, these objects were first detected only about a decade ago. Objects still more gravitationally compressed, the black holes, have become the subject of intense speculation. Black holes may have been detected in a few unusual binary systems and in the centers of some globular clusters.

Our Milky Way galaxy, too, shows signs of gravitational collapse. Initially its mass was spread throughout a giant sphere, and today the distribution of globular clusters and isolated stars is a visible vestige of our galaxy's earlier shape. Most of the original gas, dust, and stars has now been pulled into a rotating pinwheel about 100,000 light-years across. As in the formation of the solar system, rotational momentum prevented these objects from falling directly into the center of the galaxy, which is why the stars remain distributed in a flat plane rather than in a small central conglomeration.

But what about the universe as a whole? Will it not also collapse under the inexorable gravitational tug? Newton's question has been revived to become the leading problem of cosmology today. In modern terms, astronomers ask if the universe is open or closed. If space is hyperbolic, then the universe is open and unbounded, and the galaxies will forever rush away from one another, leading to an ever colder, fainter, and more tenuous distribution of matter. On the other hand, if space is spherical, then the universe is closed and bounded, and its expansion will eventually slow to a stop, followed by contraction and a mighty implosion. The open universe is a one-time affair, but the closed universe might be a single cycle of an infinite series of oscillations.

For several decades in this century the only apparent way to answer the question whether the universe is open or closed was to examine the red shifts of the distant galaxies. Hence A. R. Sandage's 1956 article, "The Red-Shift," leads off this collection, followed by G. Gamow's ("The Evolutionary Universe") and J. J. Callahan's ("The Curvature of Space in a Finite Universe") articles on the curvature of space. In all of the simple evolutionary models, the universe is gradually slowing down in its expansion; for closed space the deceleration will bring the expansion to a stop, but for open space the smaller retardation is insufficient to stop the outward rush of the galaxies. In both cases, the remote galaxies, whose light started out in the distant past, should show signs of the greater speeds at the earlier epoch. If the universe is slowing down enough to come to a stop, then these galaxies should be rushing away from us even faster than if the universe is destined to expand indefinitely. Subtle measurements are required. For very faint galaxies, which are seen at a much younger stage because they are so far away, evolutionary effects must be taken into account. Are younger galaxies intrinsically more luminous? Probably, unless separate galaxies sometimes coalesce into one as they grow older. As yet the theory is too uncertain to allow the relative distances to be established with enough precision to settle the openness or closure of the universe by deceleration observations.

If the absolute distance of the great cloud of galaxies in Virgo could be determined, then their red shifts, in combination with accurate age determinations from the relative abundances of nuclear isotopes, would help decide the question. The red-shift/distance relation ("Hubble's constant"), when directly extrapolated back to the point of zero expansion, yields an age for the universe ranging from 8 billion to 20 billion years, depending on the distance chosen for the Virgo galaxies. Because the galaxies rushed apart more rapidly in the earlier stages of the expansion, the actual age of the universe would be less than ⅔ of the directly extrapolated age if the universe were closed, and somewhat more than ⅔ if the universe were open. Thus, for example, if the direct extrapolation gave an age for the universe of 15 billion years, and an examination of the abundances of radioactive rhenium isotopes gave an age for the universe of 12 billion years, then the universe would be open, because 12 billion is greater than ⅔ of 15 billion. The difficulty lies in finding the distances to the galaxies, for the measurements have proved to be shifting and unreliable. For example, Sandage's illustration on p. 6 gives the distance of the galaxy in Virgo as 22 million light-years; in the past decade, a distance twice as great has been commonly accepted; and Sandage's most recent determination is about 70 million light-years. The other distances in the figure must be adjusted correspondingly. At present, this procedure is as unreliable as establishing the deceleration from observations of the more distant galaxies.

Another approach to the problem is to determine the mean density of the universe. If this exceeds the critical value of about 5×10^{-30} gm/cm^3, then there is sufficient gravitating mass to pull the universe back together again. Otherwise it will be Humpty Dumpty on a grand scale. (The value of the critical density depends on the expansion rate, but so does the determination of the mean density; consequently, a recalibration of intergalactic distances and the expansion rate would not affect this argument.) But ever since the 1920s, when Edwin Hubble first calculated the mean density of the universe, no one has ever found enough material to close space. Proponents of a closed universe have therefore been obliged to speculate on the whereabouts of the "missing mass."

The procedure for calculating the mean density of the universe depends on knowing the masses of various kinds of galaxies; the masses, in turn, are generally determined from the luminosities of these galaxies. However, if there is a considerable mass not accounted for by luminous objects—for example, multitudes of planets, black holes, or large gaseous extensions of the spiral disks—

then the total masses could be considerably underestimated. That this must be the case is strongly suggested by the individual motions of galaxies in clusters. These motions are sufficiently large that the galaxies in the clusters would have drifted apart unless the clusters contain invisible additional gravitating matter. Unfortunately for the advocates of a closed universe, raising the galaxy masses enough to stabilize the clusters is still not quite sufficient to raise the density of the universe to the critical value.

It is possible that the "missing mass" is in the form of hot gas between the galaxy clusters themselves. Recent calculations show that atoms heated to nearly a billion degrees Kelvin by exploding galaxies would produce X-radiation consistent with current observations. Perhaps, as Aristotle taught, nature really does abhor a vacuum, and enough highly ionized atoms will be found in intergalactic space to close the universe!

Should sufficient matter be found to raise the calculated mean density to 5×10^{-30} gm/cm^3, however, a serious conflict then would arise with predictions based on the cosmic abundance of deuterium. Astronomers presently believe that the deuterium was formed in the earliest stages of the big bang: the lower the present density of the universe, the more deuterium must have been formed at the beginning. J. R. Gott, J. E. Gunn, D. N. Schramm, and B. M. Tinsley deal cogently with this rather intricate reasoning in their article, "Will the Universe Expand Forever?" but as they admit, the details of the theory are still somewhat uncertain. Thus, although a number of arguments support the concept of an open universe, the outcome is far from settled.

Between the selections on the curvature of space and the summary arguments in favor of an open universe by Gott, Gunn, Schramm, and Tinsley there are two groups of articles. The first group concerns characteristics of the universe as a whole. D. Sciama, in "Cosmology before and after Quasars," a book review, discusses the demise of the steady-state cosmology. Then follow detailed articles on the two discoveries that brought the big-bang cosmology into ascendancy: A. Webster, "The Cosmic Background Radiation," and M. Schmidt and F. Bello, "The Evolution of Quasars." Finally, M. J. Rees and J. Silk, in "The Origin of Galaxies," relate the formation of galaxies to the evolution of the universe as a whole.

The second group deals not with the universe in the large, but with a small component whose nature may provide important insights into the structure of the universe itself. K. S. Thorne writes convincingly on "The Search for Black Holes," and S. W. Hawking, in "The Quantum Mechanics of Black Holes," describes some of the unexpected properties of bulk matter in its most compact configuration.

Like the articles on cosmology, the "+ 1" selection, "The Search for Extraterrestrial Intelligence," by C. Sagan and F. Drake, grapples with a speculative and conceivably large-scale characteristic of the universe. The possibility of detecting an extraterrestrial civilization has been wrested from science fiction and has become serious science. This is not to say that the quest finds enthusiastic support in all quarters. The chain of reasoning leading to a high probability that intelligent life exists elsewhere is no doubt even more tenuous than the cosmological dream castles of half a century ago. Nevertheless, the search is exciting and the arguments for its pursuit deserve a wider discussion. Here is a place to begin!

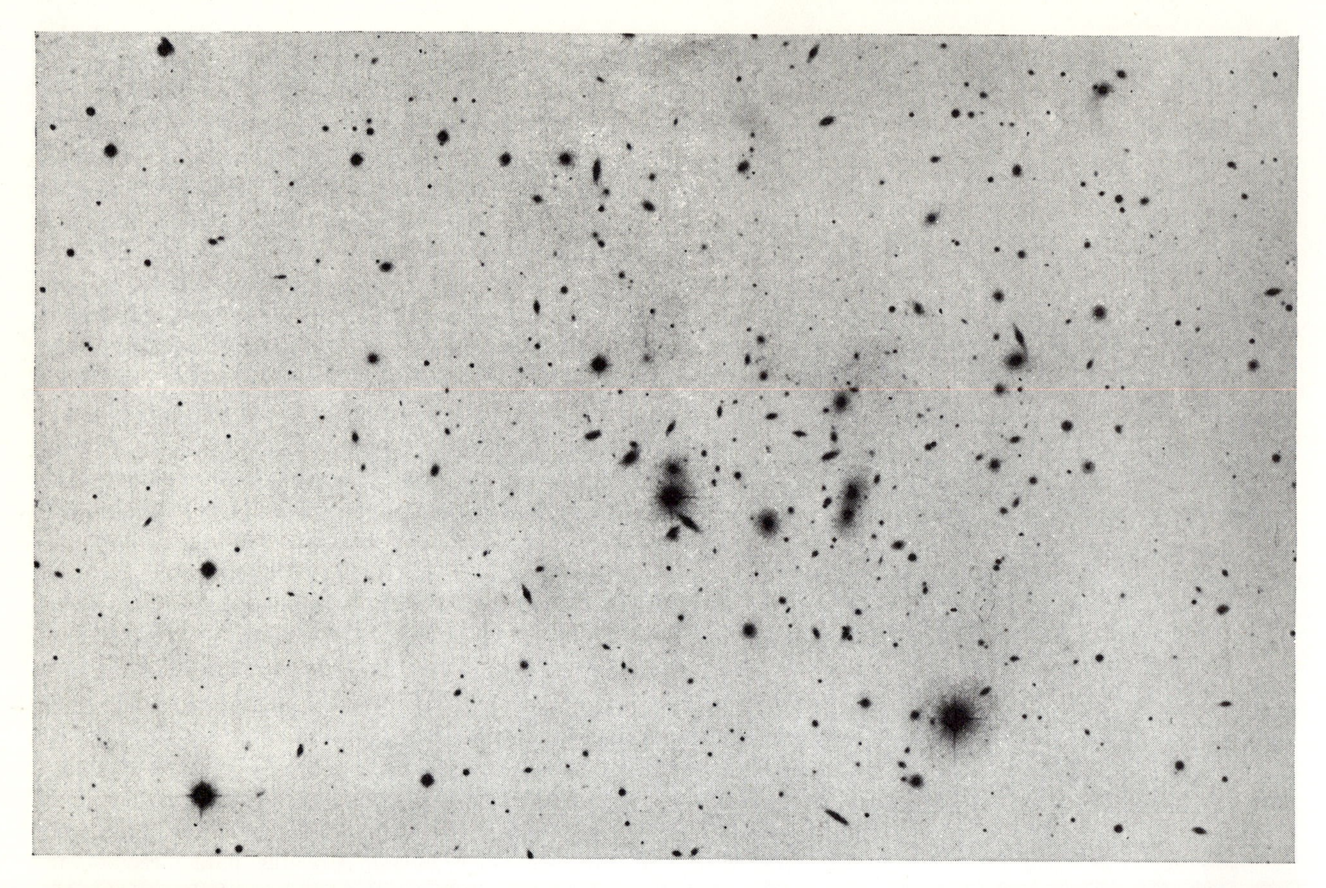

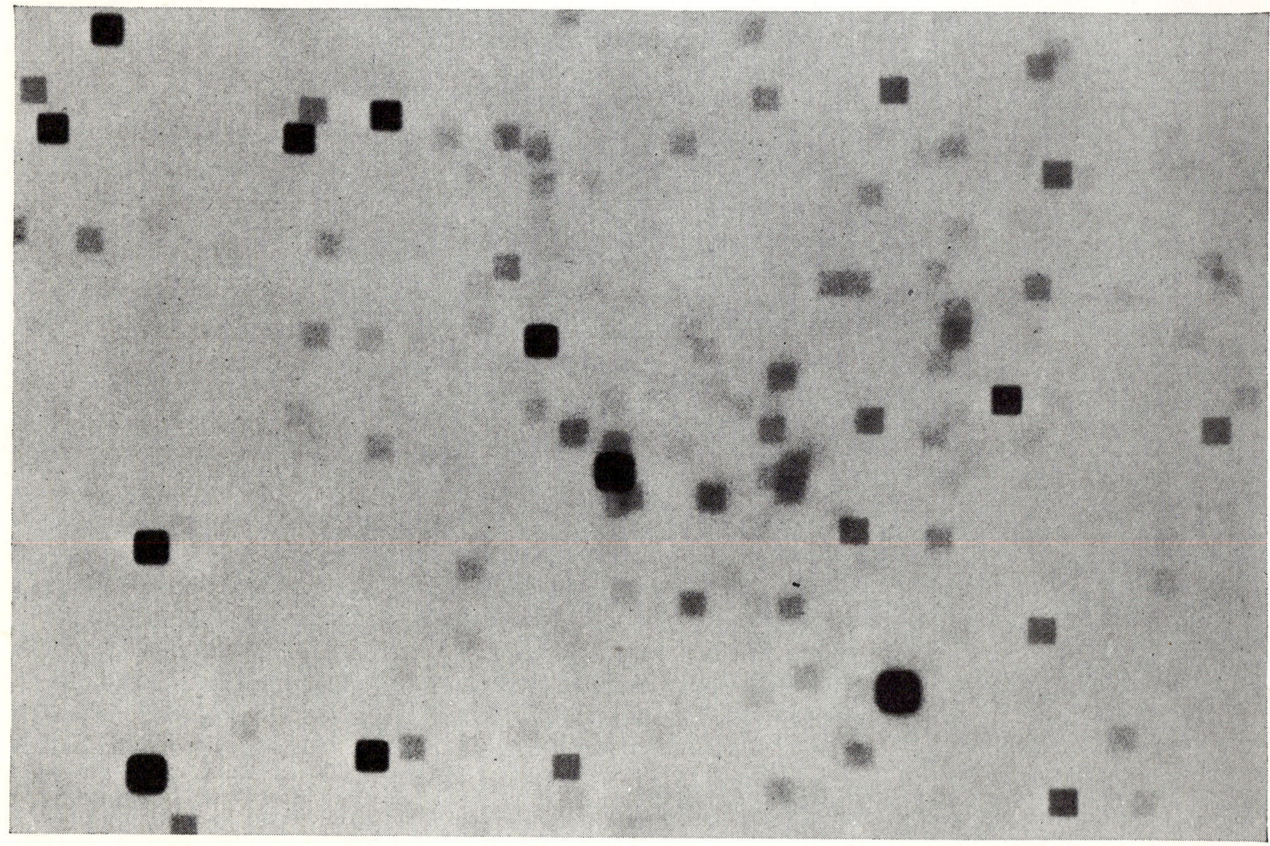

BRIGHTNESS OF GALAXIES may be measured with the help of the jiggle camera (*see photograph on the opposite page*). At the top is a negative print of a 200-inch telescope photograph showing nearby stars and a cluster of galaxies in Corona Borealis. Although the stars have made the brightest images in the photograph, they are essentially point sources of light. The galaxies, on the other hand, are extended sources of light. To measure the brightness of a galaxy by comparing it with the known brightness of a star, the two images must be made the same. This is done by smearing the images as shown at bottom in a jiggle-camera photograph of the same area.

The Red-Shift 1

by Allan R. Sandage
September 1956

The redness, and presumably the speed of recession, of most galaxies increases regularly with distance. The most distant galaxies observed appear to depart from this law, a fact of deep meaning for cosmology

In the nature of things it is a delicate undertaking to try to discern the general structure and features of a universe which stretches out farther than we can see. For more than a quarter of a century both the theoreticians and the observers of the cosmos have been making exciting discoveries, but the points of contact between the discoveries have been few. The predictions of the theorists, deduced from the most general laws of physics, are not easy to test against the real world—or rather, the small portion of the real world that we can observe. There is, however, one solid meeting ground between the theories and the observations, and that is the apparent expansion of the universe. Other aspects of the universe may be interpreted in different ways to fit different theories, but concerning the expansion the rival theories make unambiguous predictions on which they will stand or fall. There is now hope that red-shift measurements of the universe's expansion with the 200-inch telescope on Palomar Mountain will soon make it possible to decide, among other things, whether we live in an evolving or a steady-state universe.

Let us begin by considering just what issue the measurements seek to decide, as set forth by Gamow ("The Evolutionary Universe," p. 12) and Hoyle ("The Steady-State Universe, September 1956). The steady-state theory says that the universe has been expanding at a constant rate throughout an infinity of time. The evolutionary theory, in contrast, implies that the expansion of the universe is steadily slowing down. If the universe began with an explosion from a superdense state, its rate of expansion was greatest at the beginning and has been slowing ever since because of the opposing gravitational attraction of its matter, which acts as a brake on the expansion—much as an anchored elastic string attached to a golf ball would act as a brake on the flight of the ball.

Now in principle we can decide whether the rate of expansion has changed or not simply by measuring the speed of expansion at different times in the universe's history. And the 200-inch telescope permits us to do this. It covers a range of about two billion years in time. We see the nearest galaxies as they were only a few million years ago, while the light from the most distant galaxies takes so long to reach us that we see them at a stage in the universe's history going back to one or two billion years ago. If the explosion theory is correct, the universe should have been expanding at an appreciably faster rate then than it is now. Since the light we are receiving from the distant galaxies is a flashback to that earlier time, its red-shift should show them receding from us faster than if the rate of expansion had remained constant.

The red-shift is so basic a tool for testing our notions about the universe that it is worthwhile to review how it was discovered and how it is used.

An astronomer cannot perform experi-

JIGGLE CAMERA smears the images by moving the photographic plate in a rectangle during exposure. It is mounted in the prime-focus cage at the upper end of the 200-inch.

ments on the objects of his study, or even examine them at first hand. All his information rides on beams of light from outer space. By sufficiently ingenious instruments and equally ingenious interpretation (we hope), he may translate this light into information about the temperatures, sizes, structures and motions of the celestial bodies. It was in 1888 that a German astronomer, H. C. Vogel, first demonstrated that the spectra of stars could give information about motions which could not otherwise be detected. He discovered the Doppler effect in starlight.

The Doppler effect, as every physics student knows, is a change in wavelength observable when the source of radiation (sound, light, etc.) is in motion. If it is moving toward the observer, the wavelength is shortened; if away, the waves are lengthened. In the case of a star moving away from us, the whole spectrum of its light is shifted toward the red, or long-wave, end.

This spectrum, made by means of a prism or diffraction grating which spreads the light out into a band of its component colors, is usually not continuous. Certain wavelengths of the light are absorbed by atoms in the star's atmosphere. For example, most stars show strong absorption, by calcium atoms, at the wavelengths of 3933.664 and 3968.470 Angstrom units. (An Angstrom unit is a hundred-millionth of a centimeter.) The absorption is signaled by dark lines in the spectrum, known in this case as the K and H lines of calcium. Now if a star is moving away from us, these lines will be displaced toward the red end of the spectrum. In the spectrum of the star known as Delta Leporis, for in-

VIRGO

22 MILLION LIGHT-YEARS

1,200 KILOMETERS PER SECOND

CORONA BOREALIS

400 MILLION LIGHT-YEARS

21,500 KILOMETERS PER SECOND

BOOTES

700 MILLION LIGHT-YEARS

39,300 KILOMETERS PER SECOND

HYDRA

1.1 BILLION LIGHT-YEARS

60,900 KILOMETERS PER SECOND

RED-SHIFT of four galaxies on this page is depicted in the spectra on the opposite page. The galaxies are centered in the photographs. The spectra are the bright horizontal streaks tapered to the left and right. Above and below each spectrum are comparison lines from the spectrum of iron. Near the left end of the spectrum at the top of the page are two dark vertical lines: the K and H lines of calcium. If the galaxy did not exhibit the red-shift, these lines would be in the position of the broken line running vertically down the page. The amount of their shift toward the red, or right, end of the spectrum is indicated by the short arrow to the right of the broken line. The larger shift of the K and H lines of the three fainter galaxies is indicated by the longer arrows below their spectra. The constellation, approximate distance and velocity of recession of each galaxy is at left of its photograph.

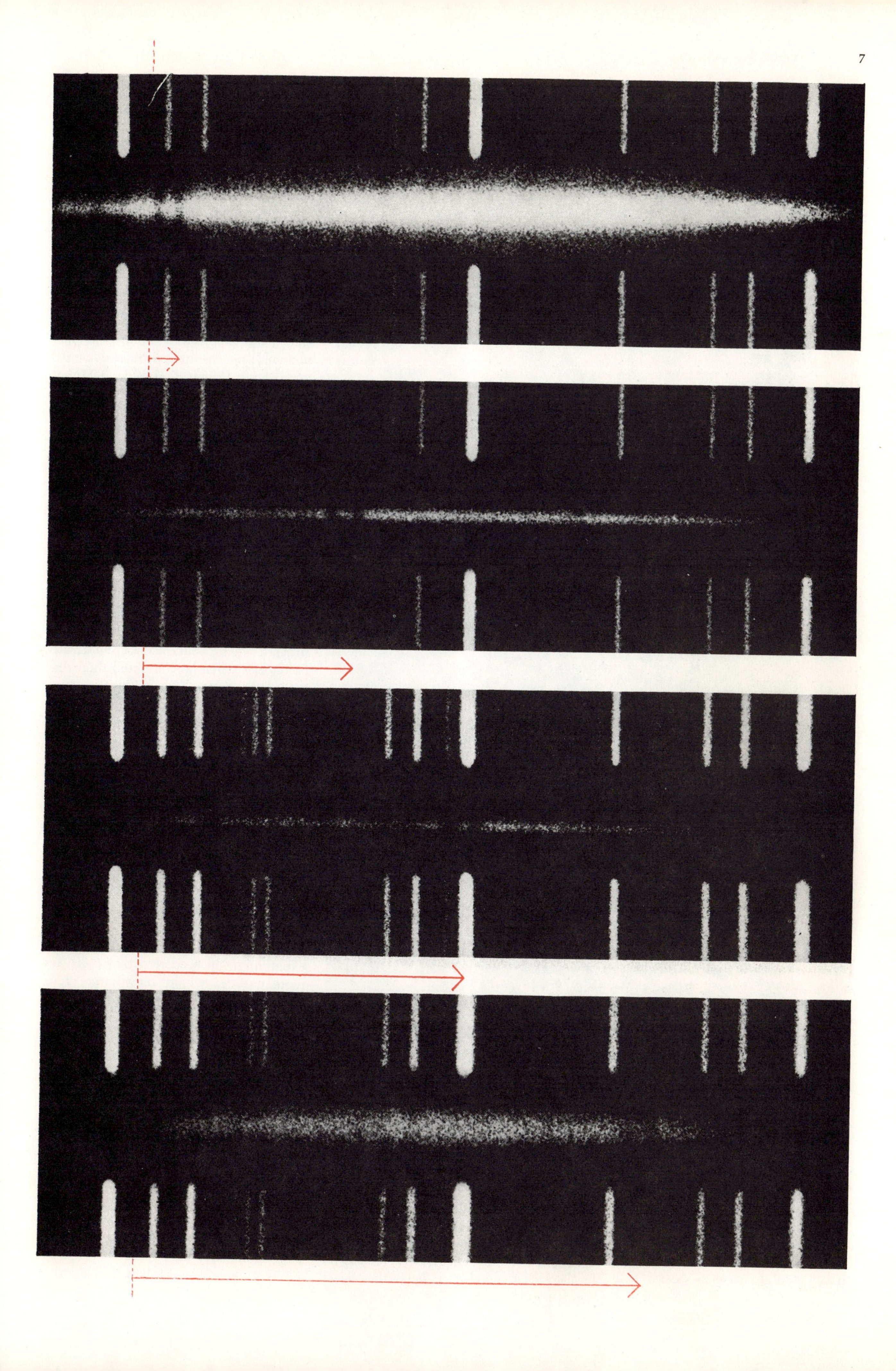

stance, the K line of calcium is displaced 1.298 Angstroms toward the red. Assuming the displacement is due to the Doppler effect, it is a simple matter to calculate the velocity of the star's receding motion. Dividing the amount of the displacement by the normal wavelength at rest, and multiplying by the speed of light (300,000 kilometers per second) we get the speed of the star—in this case 99 kilometers per second. The calculation on the basis of displacement of the H line gives the same figure.

Equipped with this powerful tool, many of the large observatories in the world spent a major part of their time during the early part of this century measuring the velocities of receding and approaching stars in our galaxy. At first it was a work of pure curiosity, no one suspecting that it might have any bearing on cosmological theories. But in the 1920s V. M. Slipher of the Lowell Observatory made a discovery which was to lead to a completely new picture of the universe. His measurements of redshifts of a number of "nebulae" then thought to lie in our galaxy showed that they were all receding from us at phenomenal speeds—up to 1,800 kilometers per second. Edwin P. Hubble at Mount Wilson soon established that the "nebulae" were systems of stars, and he went on to measure their distances. The method he used was the one developed by Harlow Shapley, employing Cepheid variable stars as the yardstick. Shapley had found a way to measure the intrinsic brightness of these stars, and therefore their distance could be estimated from their apparent brightness by means of the rule that the intensity of light falls off as the square of the distance. Hubble observed that the galaxies nearest our own system, including the Great Nebula in Andromeda, contained Cepheid variables, and when he computed their distances he came out with the then astounding figure of about one million light-years! He next tackled the problem of finding the distances of Slipher's nebulae. Since variable stars could not be detected in them, he used their brightest stars as distance indicators instead. He found that these nebulae were at distances ranging up to 20 million light-years from us, and what was more remarkable, their velocities increased in strict proportion to their distances!

SPECTRA ARE MADE with the spectrograph at the top, which is mounted at the prime focus of the 200-inch telescope. Inside the spectrograph the converging rays of the 200-inch mirror are made parallel by a concave mirror. The light is then dispersed by a diffraction grating. At the bottom is a Schmidt camera used to photograph the spectrum. It has an optical path of solid glass and a speed of $f/.48$. The plateholder and plate are below the camera.

Hubble made the daring conjecture that the universe as a whole was expanding. He predicted that more remote galaxies would show larger redshifts, still in proportion to their distance. To test Hubble's speculation, Milton L. Humason began a long-range program of spectral analysis of more distant galaxies with the 100-inch telescope on Mount Wilson. In these faint galaxies it was no longer possible to distinguish even the bright stars, and so the relative brightness of the galaxy as a whole had to be taken as the measure of distance. That is, a galaxy one fourth as bright as another was assumed to be twice as far away. Hubble reasoned that while individual galaxies might deviate from this rule, statistically the population of galaxies as a whole would follow it. The prin-

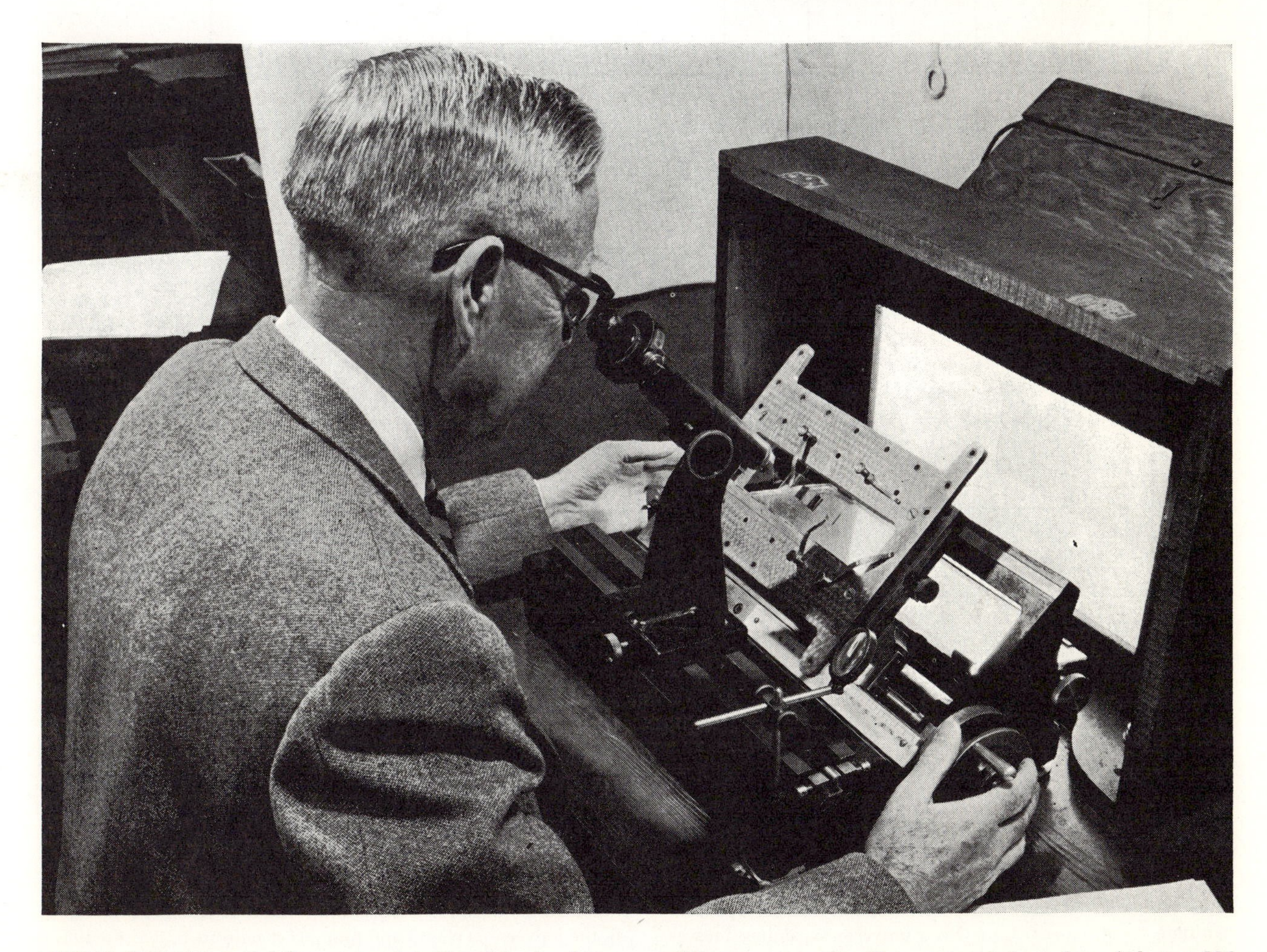

ACTUAL SIZE of the red-shift spectrum is indicated by the photograph at the top of the page. The glass photographic plate is 15 millimeters on an edge. The spectrum is 5 mm. long. At bottom Milton L. Humason examines a spectrum with a low-power microscope.

ciple is still the basis of distance determinations today.

Humason laboriously photographed spectra of galaxies, and Hubble measured their apparent brightness, from 1928 to 1936, when they reached the limit of the 100-inch telescope. The history of the red-shift program in those years is a story of extreme skill and patience at the telescope and of steady improvement in instrumentation. It was a long and difficult task to photograph spectra then; the prisms used required long exposures, and it took 10 nights or more to obtain a spectrum which with modern equipment can be recorded in less than an hour today. The improvement in equipment includes not only the 200-inch telescope but also diffraction gratings, faster cameras and a vast improvement in the sensitivity of photographic plates, thanks to the Eastman Kodak Company. Astronomers the world over, and cosmology, owe a large debt to the Eastman research laboratories.

Humason's first really big red-shift came early in 1928, when he got a spectrum of a galaxy called NGC 7619. Hubble had predicted that its velocity should be slightly less than 4,000 kilometers per second: Humason found it to be 3,800. By 1936, at the limit of the 100-inch telescope's reach, they had arrived at a cluster of galaxies, called Ursa Major No. 2, which showed a velocity of 40,000 kilometers per second. All the way out to that range of more than half a billion light-years the velocity of galaxies increased in direct proportion to the distance. In a sense this was disappointing, because the various cosmological theories predicted that some change in this relation should begin to appear when the observations had been pushed far enough. Further exploration into the distances of space had to await the completion of the 200-inch Hale telescope on Palomar Mountain.

In 1951 the red-shift program was resumed, with a new spectrograph of great speed and versatility placed in the big telescope's prime focus cage, where the observer rides with his instruments. The spectrograph has to be of very compact design to fit into the cramped space of the cage. The photographic plate itself, mounted in the middle of a complex

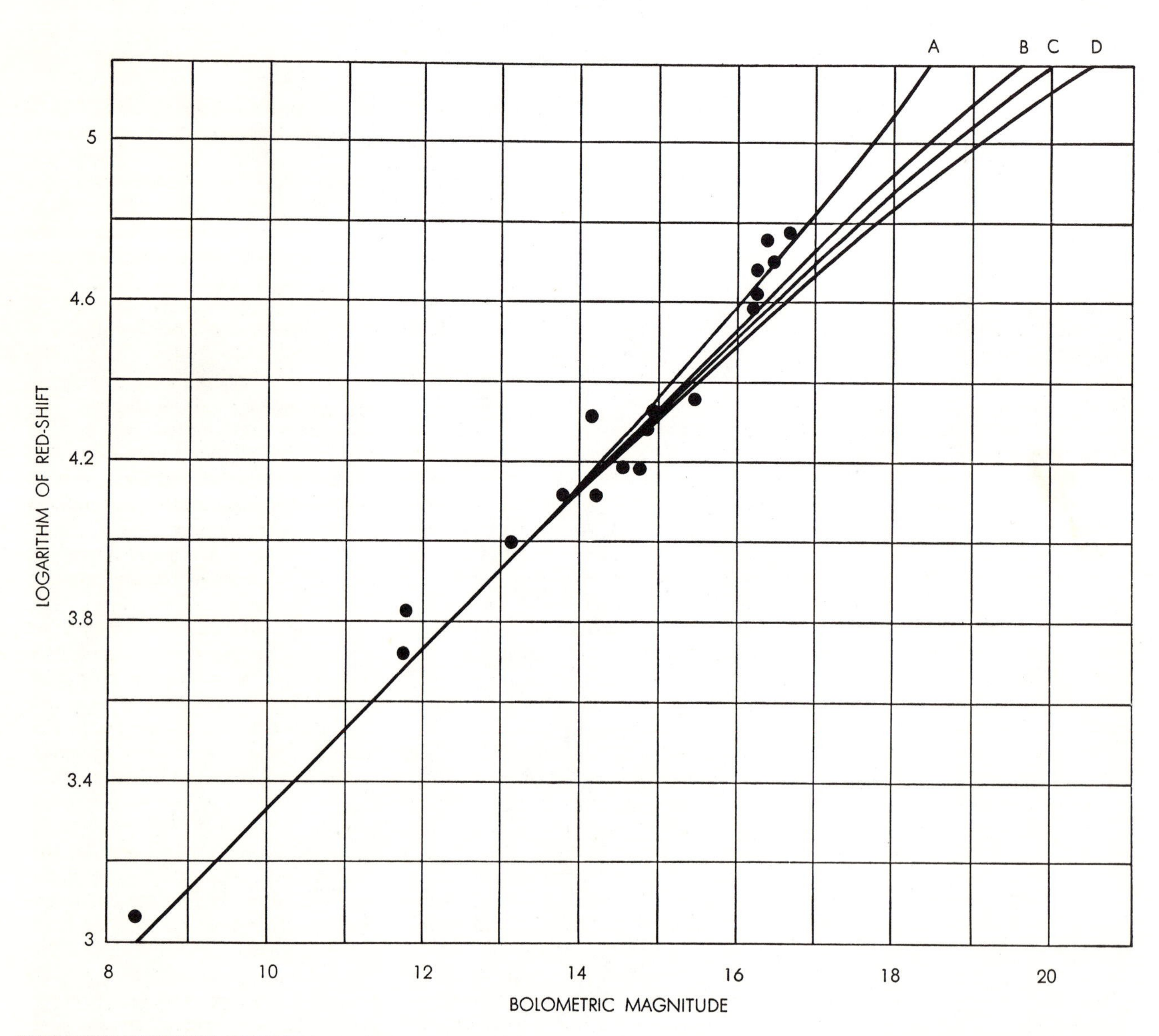

EIGHTEEN FAINTEST CLUSTERS of galaxies yet measured are plotted for their red-shift (or speed of recession) and apparent magnitude (or distance). Line C is a universe expanding forever at the same rate. Line D is a steady-state universe. If the line falls to the left of C, the expansion must slow down. If it falls between C and B, the universe is open and infinite. If it falls to the left of B, the universe is closed and finite. If it falls on B, it is Euclidean and infinite. A is the trend suggested by the six faintest clusters.

optical arrangement, is only 15 millimeters (about half an inch) on a side. The cutting and handling of such small pieces of glass in complete darkness (to avoid exposure of the plate) is a tricky business. The spectrum recorded on the plate is a tiny strip only a fifth of an inch long, but it is long enough to measure red-shifts to an accuracy of better than one half of 1 per cent.

The most distant photographable galaxies are so faint that they are not visible to the eye through the telescope: they can be recorded only by extended exposure of the plate. The observer guiding the telescope must position the slit of the spectrograph by reference to guide stars within the same field as the distant object. Another great difficulty in recording the red-shift of extremely distant galaxies arises from the magnitude of the shift. The displacement of the calcium dark lines toward the red is so large that the lines move clean off the sensitive range of blue photographic plates, which astronomers like to use because of their speed. So slow panchromatic plates must be used, and Humason has been forced to return to exposure times as long as 30 hours or more.

The other part of the program—measuring the distances of the galaxies—also has been helped by improvements in technique. For measurement of their brightness the Mount Wilson telescopes employ photomultiplier tubes, which amplify the light energy by electronic means. Such equipment was not available for the 200-inch telescope when the present program began. Instead the intensity of the light from very faint galaxies was measured by a tricky method which compares it with that of stars of known magnitude. No direct comparison can be made, of course, between the picture of a star and that of a galaxy or cluster of galaxies, because the star is a point source of light while a galactic system is a spread-out image. To make the images comparable, a region of the sky is photographed with a "jiggle" camera which moves the plate around so that the images of stars and of galaxies are smeared out in squares [*see photograph at bottom of page 4*]. They can then be compared as to brightness—just as one may use color cards to find a match to the color of a room.

Humason has now measured red-shifts of remote clusters of galaxies with recession velocities up to 60,000 kilometers per second. What do they show? Is the velocity still increasing in strict proportion to the distance?

The information about 18 of the faintest measured clusters is given in the accompanying chart [*see page 10*]. Their velocities are plotted against their apparent brightness, or estimated distances. If velocity increases in direct proportion to the distance, the observed velocity-distance relation should be "linear" (*i.e.*, follow a straight line). But as the chart shows, the very faintest clusters have begun to depart from that line. These clusters, about a billion light-years away, are moving *faster* (by about 10,000 kilometers per second) than in direct proportion to their apparent distance. In other words, the data would be interpreted to mean that a billion years ago the universe was expanding faster than it is now. If the measurements and the interpretation are correct, this suggests that we live in an evolving rather than in a steady-state universe.

The observed change in the curve buys us much more information. To begin with, it tells us something about the mean density of matter in the universe. The rate at which the expansion of the universe is slowing down (if it is) depends on the mean density of its matter: the higher the density, the greater the braking effect. The amount of departure from linearity indicated by the measurements thus far calls for a mean density of about 3×10^{-28} grams of matter per cubic centimeter (about one hydrogen atom per five quarts of space). Now this amounts to about 300 times the total mass of the matter estimated to be contained in galaxies: that figure comes out to a mean density of only 10^{-30} grams per cubic centimeter. If our present tentative value for the slowdown of the expansion should be confirmed, we would have to conclude that either the current estimates of the masses of the galaxies are wrong or that there is a great deal of matter, so far undetected, in intergalactic space. Matter in the form of neutral hydrogen (*i.e.*, normal hydrogen atoms consisting of a proton and an electron) might be present in space and still have escaped detection until now because it is not luminous. The giant radio telescopes now under construction or on the drawing boards perhaps will detect the hydrogen, if it exists in the postulated quantities.

Once we know the rate at which expansion of the universe is slowing down, it becomes possible to determine not only the mean density of matter but also the geometry of space—that is, its curvature. Models of the evolving universe take three forms: the Euclidean case, in which space is flat, open and infinite; a curved universe which is closed and finite, like the surface of a sphere; and a curved universe which is open and infinite, like the surface of a saddle. In the accompanying velocity-distance chart [*page 10*] curves to the left of C represent evolving models, and curve D represents the steady-state model. If the curve of the velocity-distance relation lies between C and B, the universe is open and infinite. Line B is the Euclidean case of flat space. If the curve is left of B, the universe is closed and finite, the radius of its curvature decreasing as we move farther to the left.

According to our present observations, the actual relation follows a curve left of B (curve A on the chart). Although our data are still crude and inconclusive, they do suggest that the steady-state model does not fit the real world, and that we live in a closed, evolving universe.

Humason has gone beyond 60,000 kilometers per second and attempted to measure the red-shifts of two faint clusters whose predicted velocity is more than 100,000 kilometers per second. So far these efforts have not yielded reliable results, but he is continuing them. These two remote clusters may well hold the key to the structure of the universe. We stand a chance of finding the answer to the cosmological problem. The red-shift program will continue toward this goal.

If the expansion of the universe is decelerating at the rate our present data suggest, the expansion will eventually stop and contraction will begin. If it returns to a superdense state and explodes again, then in the next cycle of oscillation, some 15 billion years hence, we may all find ourselves again pursuing our present tasks.

Although no final answers have yet emerged, big steps have been taken since 1928 toward the solution to the cosmological problem, and there is hope that it may now be within our grasp. The situation has nowhere been better expressed than in Hubble's last paper:

"For I can end as I began. From our home on the earth we look out into the distances and strive to imagine the sort of world into which we are born. Today we have reached far out into space. Our immediate neighborhood we know rather intimately. But with increasing distance our knowledge fades . . . until at the last dim horizon we search among ghostly errors of observations for landmarks that are scarcely more substantial. The search will continue. The urge is older than history. It is not satisfied and it will not be suppressed."

2 The Evolutionary Universe

by George Gamow
September 1956

Most cosmologists believe that the universe began as a dense kernel of matter and radiant energy which started to expand about five billion years ago and later coalesced into galaxies

Cosmology is the study of the general nature of the universe in space and in time—what it is now, what it was in the past and what it is likely to be in the future. Since the only forces at work between the galaxies that make up the material universe are the forces of gravity, the cosmological problem is closely connected with the theory of gravitation, in particular with its modern version as comprised in Albert Einstein's general theory of relativity. In the frame of this theory the properties of space, time and gravitation are merged into one harmonious and elegant picture.

The basic cosmological notion of general relativity grew out of the work of great mathematicians of the 19th century. In the middle of the last century two inquisitive mathematical minds—a Russian named Nikolai Lobachevski and a Hungarian named János Bolyai—discovered that the classical geometry of Euclid was not the only possible geometry: in fact, they succeeded in constructing a geometry which was fully as logical and self-consistent as the Euclidean. They began by overthrowing Euclid's axiom about parallel lines: namely, that only one parallel to a given straight line can be drawn through a point not on that line. Lobachevski and Bolyai both conceived a system of geometry in which a great number of lines parallel to a given line could be drawn through a point outside the line.

To illustrate the differences between Euclidean geometry and their non-Euclidean system it is simplest to consider just two dimensions—that is, the geometry of surfaces. In our schoolbooks this is known as "plane geometry," because the Euclidean surface is a flat surface. Suppose, now, we examine the properties of a two-dimensional geometry constructed not on a plane surface but on a curved surface. For the system of Lobachevski and Bolyai we must take the curvature of the surface to be "negative," which means that the curvature is not like that of the surface of a sphere but like that of a saddle [*see illustrations on page 14*]. Now if we are to draw parallel lines or any figure (*e.g.*, a triangle) on this surface, we must decide first of all how we shall define a "straight line," equivalent to the straight line of plane geometry. The most reasonable definition of a straight line in Euclidean geometry is that it is the path of the shortest distance between two points. On a curved surface the line, so defined, becomes a curved line known as a "geodesic" [see "The Straight Line," by Morris Kline; SCIENTIFIC AMERICAN, March 1956].

Considering a surface curved like a saddle, we find that, given a "straight" line or geodesic, we can draw through a point outside that line a great many geodesics which will never intersect the given line, no matter how far they are extended. They are therefore parallel to it, by the definition of parallel. The possible parallels to the line fall within certain limits, indicated by the intersecting

Five contributors to modern cosmology are depicted in these drawings by Bernarda Bryson.

lines in the drawing at the left in the middle of the next page.

As a consequence of the overthrow of Euclid's axiom on parallel lines, many of his theorems are demolished in the new geometry. For example, the Euclidean theorem that the sum of the three angles of a triangle is 180 degrees no longer holds on a curved surface. On the saddle-shaped surface the angles of a triangle formed by three geodesics always add up to less than 180 degrees, the actual sum depending on the size of the triangle. Further, a circle on the saddle surface does not have the same properties as a circle in plane geometry. On a flat surface the circumference of a circle increases in proportion to the increase in diameter, and the area of a circle increases in proportion to the square of the increase in diameter. But on a saddle surface both the circumference and the area of a circle increase at *faster* rates than on a flat surface with increasing diameter.

After Lobachevski and Bolyai, the German mathematician Bernhard Riemann constructed another non-Euclidean geometry whose two-dimensional model is a surface of positive, rather than negative, curvature—that is, the surface of a sphere. In this case a geodesic line is simply a great circle around the sphere or a segment of such a circle, and since any two great circles must intersect at two points (the poles), there are no parallel lines at all in this geometry. Again the sum of the three angles of a triangle is not 180 degrees: in this case it is always *more* than 180. The circumference of a circle now increases at a rate *slower* than in proportion to its increase in diameter, and its area increases more slowly than the square of the diameter.

Now all this is not merely an exercise in abstract reasoning but bears directly on the geometry of the universe in which we live. Is the space of our universe "flat," as Euclid assumed, or is it curved negatively (per Lobachevski and Bolyai) or curved positively (Riemann)? If we were two-dimensional creatures living in a two-dimensional universe, we could tell whether we were living on a flat or a curved surface by studying the properties of triangles and circles drawn on that surface. Similarly as three-dimensional beings living in three-dimensional space we should be able, by studying geometrical properties of that space, to decide what the curvature of our space is. Riemann in fact developed mathematical formulas describing the properties of various kinds of curved space in three and more dimensions. In the early years of this century Einstein conceived the idea of the universe as a curved system in four dimensions, embodying time as the fourth dimension, and he proceeded to apply Riemann's formulas to test his idea.

Einstein showed that time can be considered a fourth coordinate supplementing the three coordinates of space. He connected space and time, thus establishing a "space-time continuum," by means of the speed of light as a link between time and space dimensions. However, recognizing that space and time are physically different entities, he employed the imaginary number $\sqrt{-1}$, or i, to express the unit of time mathematically and make the time coordinate formally equivalent to the three coordinates of space.

In his special theory of relativity Einstein made the geometry of the time-space continuum strictly Euclidean, that is, flat. The great idea that he introduced later in his general theory was that gravitation, whose effects had been neglected in the special theory, must make it curved. He saw that the gravitational effect of the masses distributed in space and moving in time was equivalent to curvature of the four-dimensional space-time continuum. In place of the classical Newtonian statement that "the sun produces a field of forces which impels the earth to deviate from straight-line mo-

From left to right they are: Nikolai Lobachevski, Bernhard Riemann, Albert Einstein, Willem de Sitter and Georges Lemaître

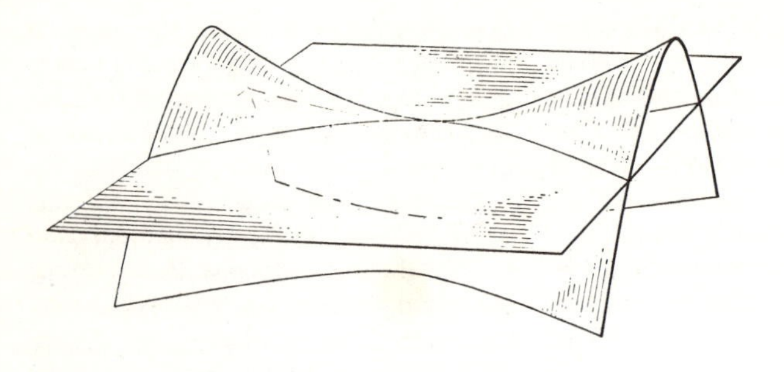

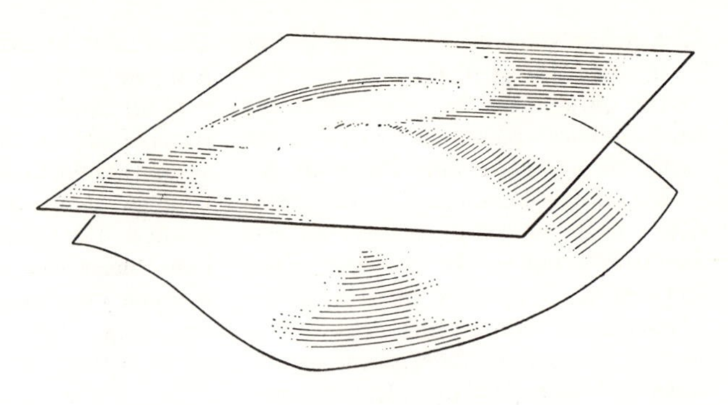

NEGATIVE AND POSITIVE CURVATURE of space is suggested by this two-dimensional analogy. The saddle-shaped surface at left, which lies on both sides of a tangential plane, is negatively curved. The spherical surface at right, which lies on one side of a tangential plane, is positively curved. If space is negatively curved, the universe is infinite; if it is positively curved, the universe is finite.

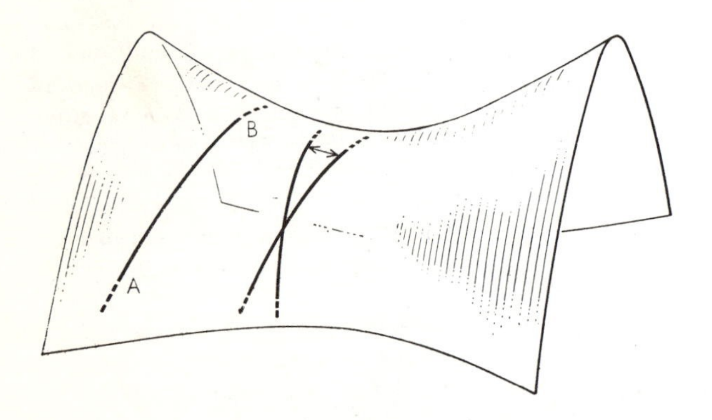

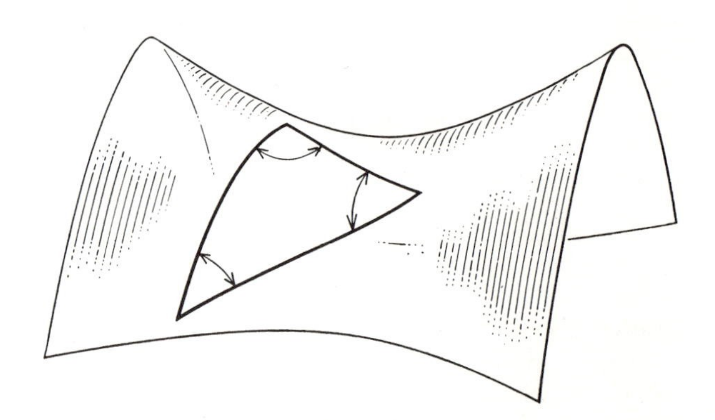

ON A NEGATIVELY CURVED SURFACE the shortest distance between two points is not a straight line but a curved "geodesic," such as the line AB at the left. On a plane surface only one parallel to a given straight line can be drawn through a point not on that line; on a negatively curved surface many geodesics can be drawn through a point not on a given geodesic without ever intersecting it. These "parallel" lines will fall within the limits indicated by the arrow between the intersecting lines at left. On a plane surface the angles of a triangle add up to 180 degrees; on the negatively curved surface at the right, they add up to less than 180 degrees.

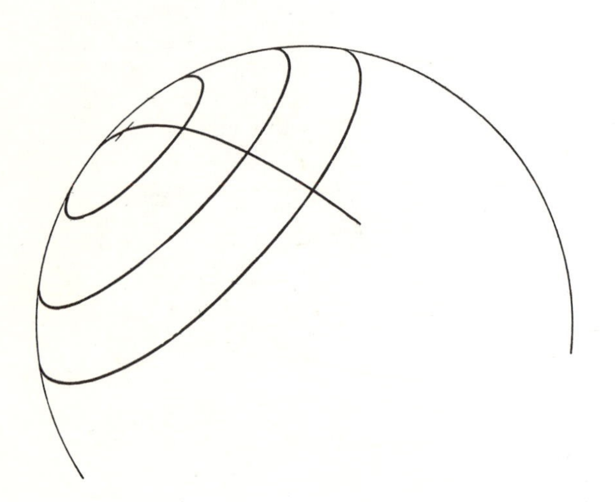

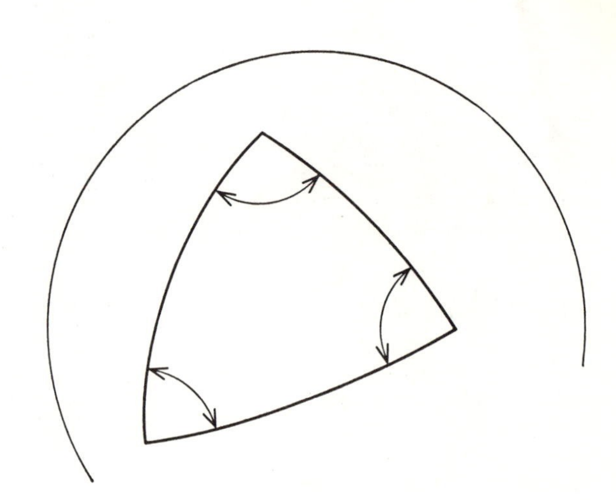

ON A POSITIVELY CURVED SURFACE the shortest distance between two points follows a great circle, a closed line passing through opposite points on the surface (*single curved line at left*). In this geometry there are no "parallel" lines because any two great circles must intersect. The circumference of a circle increases more slowly with diameter than on a flat surface, and the area similarly increases more slowly (*concentric circles at left*). The angles of a triangle on the surface (*right*) add up to more than 180 degrees.

tion and to move in a circle around the sun," Einstein substituted a statement to the effect that "the presence of the sun causes a curvature of the space-time continuum in its neighborhood."

The motion of an object in the space-time continuum can be represented by a curve called the object's "world line." For example, the world line of the earth's travel around the sun in time is pictured in the drawing on this page. (Space must be represented here in only two dimensions; it would be impossible for a three-dimensional artist to draw the fourth dimension in this scheme, but since the orbit of the earth around the sun lies in a single plane, the omission is unimportant.) Einstein declared, in effect: "The world line of the earth is a geodesic in the curved four-dimensional space around the sun." In other words, the line ABCD in the drawing corresponds to the shortest *four-dimensional* distance between the position of the earth in January (at A) and its position in October (at D).

Einstein's idea of the gravitational curvature of space-time was, of course, triumphantly affirmed by the discovery of perturbations in the motion of Mercury at its closest approach to the sun and of the deflection of light rays by the sun's gravitational field. Einstein next attempted to apply the idea to the universe as a whole. Does it have a general curvature, similar to the local curvature in the sun's gravitational field? He now had to consider not a single center of gravitational force but countless centers of attraction in a universe full of matter concentrated in galaxies whose distribution fluctuates considerably from region to region in space. However, in the large-scale view the galaxies are spread fairly uniformly throughout space as far out as our biggest telescopes can see, and we can justifiably "smooth out" its matter to a general average (which comes to about one hydrogen atom per cubic meter). On this assumption the universe as a whole has a smooth general curvature.

But if the space of the universe is curved, what is the sign of this curvature? Is it positive, as in our two-dimensional analogy of the surface of a sphere, or is it negative, as in the case of a saddle surface? And, since we cannot consider space alone, how is this space curvature related to time?

Analyzing the pertinent mathematical equations, Einstein came to the conclusion that the curvature of space must be independent of time, *i.e.*, that the universe as a whole must be unchanging (though it changes internally). However, he found to his surprise that there was no solution of the equations that would permit a static cosmos. To repair the situation, Einstein was forced to introduce an additional hypothesis which amounted to the assumption that a new kind of force was acting among the galaxies. This hypothetical force had to be independent of mass (being the same for an apple, the moon and the sun!) and to gain in strength with increasing distance between the interacting objects (as no other forces ever do in physics!).

Einstein's new force, called "cosmic repulsion," allowed two mathematical models of a static universe. One solution, which was worked out by Einstein him-

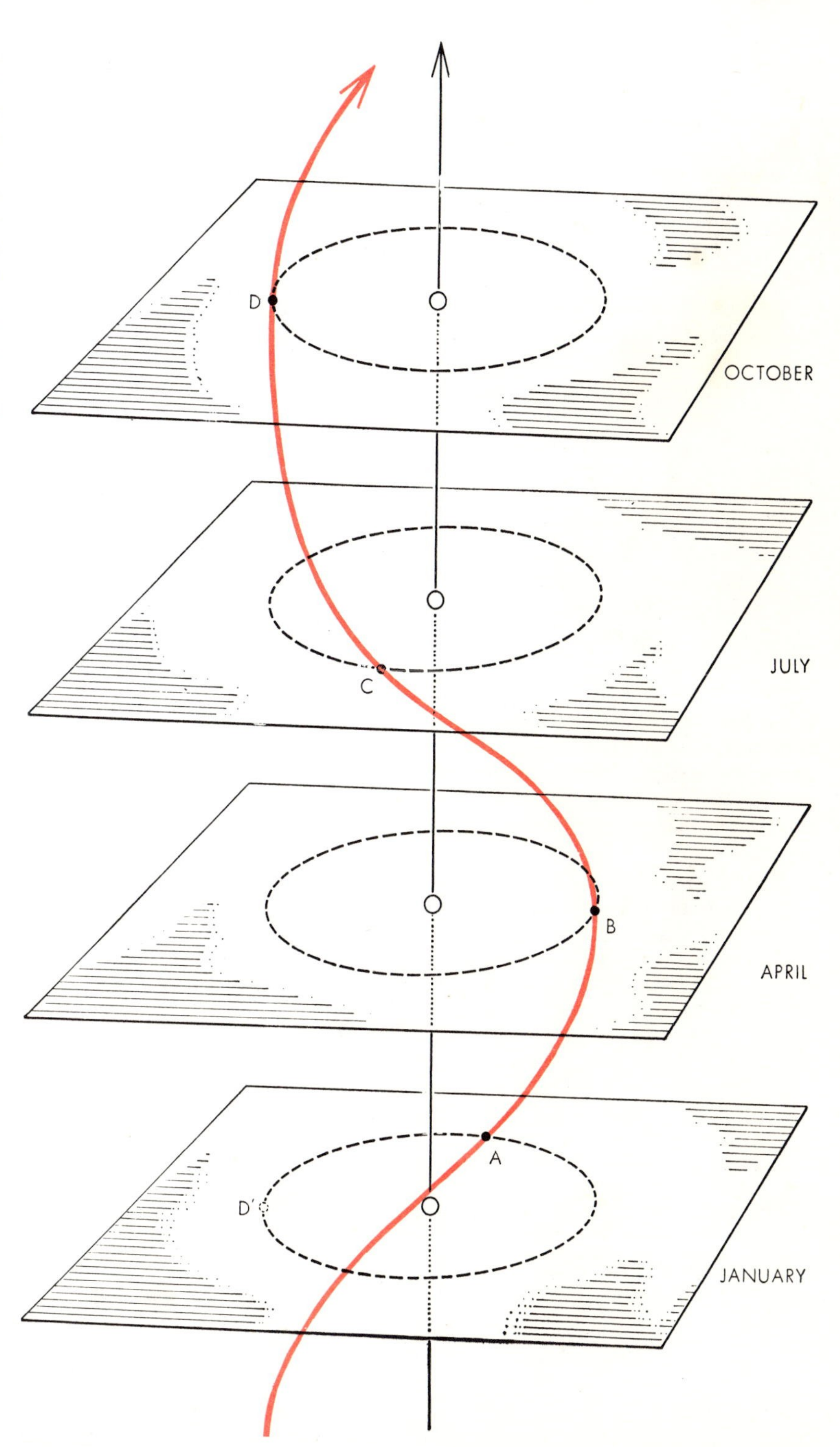

MOTION OF BODY in the curved "space-time continuum" of Albert Einstein is represented by the "world line" of the earth's motion around the sun. Here the sun is the small open circle in each of the four planes. The earth is the black dot on the elliptical orbit. Each plane shows the position of the earth at a month of the year. The world line is in color.

self and became known as "Einstein's spherical universe," gave the space of the cosmos a positive curvature. Like a sphere, this universe was closed and thus had a finite volume. The space coordinates in Einstein's spherical universe were curved in the same way as the latitude or longitude coordinates on the surface of the earth. However, the time axis of the space-time continuum ran quite straight, as in the good old classical physics. This means that no cosmic event would ever recur. The two-dimensional analogy of Einstein's space-time continuum is the surface of a cylinder, with the time axis running parallel to the axis of the cylinder and the space axis perpendicular to it [*see drawing at left on this page*].

The other static solution based on the mysterious repulsion forces was discovered by the Dutch mathematician Willem de Sitter. In his model of the universe both space and time were curved. Its geometry was similar to that of a globe, with longitude serving as the space coordinate and latitude as time [*drawing at right on this page*].

Unhappily astronomical observations contradicted both Einstein's and de Sitter's static models of the universe, and they were soon abandoned.

In the year 1922 a major turning point came in the cosmological problem. A Russian mathematician, Alexander A. Friedman (from whom the author of this article learned his relativity), discovered an error in Einstein's proof for a static universe. In carrying out his proof Einstein had divided both sides of an equation by a quantity which, Friedman found, could become zero under certain circumstances. Since division by zero is not permitted in algebraic computations, the possibility of a nonstatic universe could not be excluded under the circumstances in question. Friedman showed that two nonstatic models were possible. One pictured the universe as expanding with time; the other, contracting.

Einstein quickly recognized the importance of this discovery. In the last edition of his book *The Meaning of Relativity* he wrote: "The mathematician Friedman found a way out of this dilemma. He showed that it is possible, according to the field equations, to have a finite density in the whole (three-dimensional) space, without enlarging these field equations ad hoc." Einstein remarked to me many years ago that the cosmic repulsion idea was the biggest blunder he had made in his entire life.

Almost at the very moment that Friedman was discovering the possibility of an expanding universe by mathematical reasoning, Edwin P. Hubble at the Mount Wilson Observatory on the other side of the world found the first evidence of actual physical expansion through his telescope. He made a compilation of the distances of a number of far galaxies, whose light was shifted toward the red end of the spectrum, and it was soon found that the extent of the shift was in direct proportion to a galaxy's distance from us, as estimated by its faintness. Hubble and others interpreted the red-shift as the Doppler effect—the well-known phenomenon of lengthening of wavelengths from any radiating source that is moving rapidly away (a train whistle, a source of light or whatever). To date there has been no other reasonable explanation of the galaxies' red-shift. If the explanation is correct, it means that the galaxies are all moving away from one another with increasing velocity as they move farther apart.

Thus Friedman and Hubble laid the foundation for the theory of the expanding universe. The theory was soon developed further by a Belgian theoretical astronomer, Georges Lemaître. He proposed that our universe started from a highly compressed and extremely hot state which he called the "primeval atom." (Modern physicists would prefer the term "primeval nucleus.") As this matter expanded, it gradually thinned out, cooled down and reaggregated in stars and galaxies, giving rise to the highly complex structure of the universe as we know it today.

Until a few years ago the theory of the expanding universe lay under the cloud of a very serious contradiction. The measurements of the speed of flight of the galaxies and their distances from us indicated that the expansion had started about 1.8 billion years ago. On the other hand, measurements of the age of ancient rocks in the earth by the clock of radioactivity (*i.e.*, the decay of uranium to lead) showed that some of the rocks were at least three billion years old; more recent estimates based on other radioactive elements raise the age of the earth's crust to almost five billion years. Clearly a universe 1.8 billion years old could not contain five-billion-year-old rocks! Happily the contradiction has now been disposed of by Walter Baade's recent discovery that the distance yardstick (based on the periods of variable stars) was faulty and that the distances between galaxies are more than twice as great as they were thought to be. This change in distances raises the age of the universe to five billion years or more.

Friedman's solution of Einstein's cosmological equation, as I mentioned, permits two kinds of universe. We can call one the "pulsating" universe. This model says that when the universe has reached a certain maximum permissible expansion, it will begin to contract; that it will shrink until its matter has been compressed to a certain maximum density, possibly that of atomic nuclear material, which is a hundred million million times denser than water; that it will then begin to expand again—and so on through the cycle *ad infinitum*. The other model is a "hyperbolic" one: it suggests that from an infinitely thin state an eternity ago the universe contracted until it reached the maximum density, from which it rebounded to an unlimited expansion which will go on indefinitely in the future.

The question whether our universe is actually "pulsating" or "hyperbolic" should be decidable from the present rate of its expansion. The situation is analogous to the case of a rocket shot from the surface of the earth. If the velocity of the rocket is less than seven miles per second—the "escape velocity"—the rocket will climb only to a certain

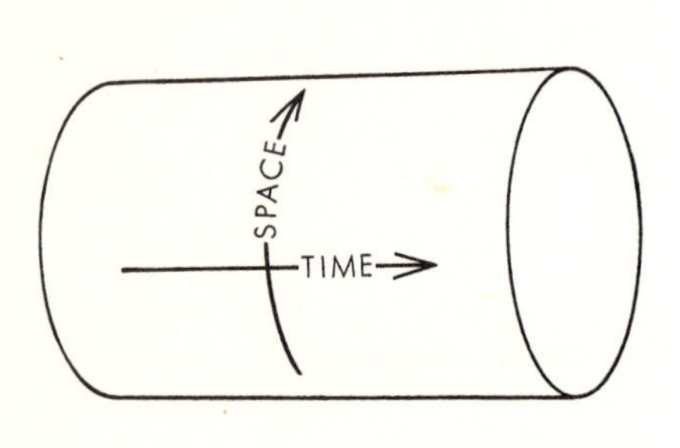

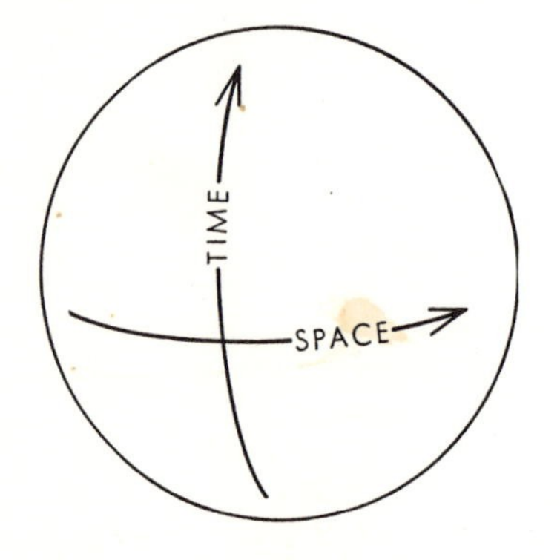

SPHERICAL UNIVERSE of Einstein may be represented in two dimensions by a cylinder (*left*). Its space coordinates were positively curved but its time coordinate was straight. The spherical universe of Willem de Sitter had positively curved coordinates (*right*).

height and then fall back to the earth. (If it were completely elastic, it would bounce up again, etc., etc.) On the other hand, a rocket shot with a velocity of more than seven miles per second will escape from the earth's gravitational field and disappear in space. The case of the receding system of galaxies is very similar to that of an escape rocket, except that instead of just two interacting bodies (the rocket and the earth) we have an unlimited number of them escaping from one another. We find that the galaxies are fleeing from one another at seven times the velocity necessary for mutual escape.

Thus we may conclude that our universe corresponds to the "hyperbolic" model, so that its present expansion will never stop. We must make one reservation. The estimate of the necessary escape velocity is based on the assumption that practically all the mass of the universe is concentrated in galaxies. If intergalactic space contained matter whose total mass was more than seven times that in the galaxies, we would have to reverse our conclusion and decide that the universe is pulsating. There has been no indication so far, however, that any matter exists in intergalactic space, and it could have escaped detection only if it were in the form of pure hydrogen gas, without other gases or dust.

Is the universe finite or infinite? This resolves itself into the question: Is the curvature of space positive or negative—closed like that of a sphere, or open like that of a saddle? We can look for the answer by studying the geometrical properties of its three-dimensional space, just as we examined the properties of figures on two-dimensional surfaces. The most convenient property to investigate astronomically is the relation between the volume of a sphere and its radius.

We saw that, in the two-dimensional case, the area of a circle increases with increasing radius at a faster rate on a negatively curved surface than on a Euclidean or flat surface; and that on a positively curved surface the relative rate of increase is slower. Similarly the increase of volume is faster in negatively curved space, slower in positively curved space. In Euclidean space the volume of a sphere would increase in proportion to the cube, or third power, of the increase in radius. In negatively curved space the volume would increase faster than this; in positively curved space, slower. Thus if we look into space and find that the volume of successively larger spheres, as measured by a count of the galaxies within them, increases faster than the cube of the distance to the limit of the sphere (the radius), we can conclude that the space of our universe has negative curvature, and therefore is open and infinite. By the same token, if the number of galaxies increases at a rate slower than the cube of the distance, we live in a universe of positive curvature—closed and finite.

Following this idea, Hubble undertook to study the increase in number of galaxies with distance. He estimated the distances of the remote galaxies by their relative faintness: galaxies vary considerably in intrinsic brightness, but over a very large number of galaxies these variations are expected to average out. Hubble's calculations produced the conclusion that the universe is a closed system—a small universe only a few billion light-years in radius!

We know now that the scale he was using was wrong: with the new yardstick the universe would be more than twice as large as he calculated. But there is a more fundamental doubt about his result. The whole method is based on the assumption that the intrinsic brightness of a galaxy remains constant. What if it changes with time? We are seeing the light of the distant galaxies as it was emitted at widely different times in the past—500 million, a billion, two billion years ago. If the stars in the galaxies are burning out, the galaxies must dim as they grow older. A galaxy two billion light-years away cannot be put on the same distance scale with a galaxy 500 million light-years away unless we take into account the fact that we are seeing the nearer galaxy at an older, and less bright, age. The remote galaxy is farther away than a mere comparison of the luminosity of the two would suggest.

When a correction is made for the assumed decline in brightness with age, the more distant galaxies are spread out to farther distances than Hubble assumed. In fact, the calculations of volume are changed so drastically that we may have to reverse the conclusion about the curvature of space. We are not sure, because we do not yet know enough about the evolution of galaxies. But if we find that galaxies wane in intrinsic brightness by only a few per cent in a billion years, we shall have to conclude that space is curved negatively and the universe is infinite.

Actually there is another line of reasoning which supports the side of infinity. Our universe seems to be hyperbolic and ever-expanding. Mathematical solutions of fundamental cosmological equations indicate that such a universe is open and infinite.

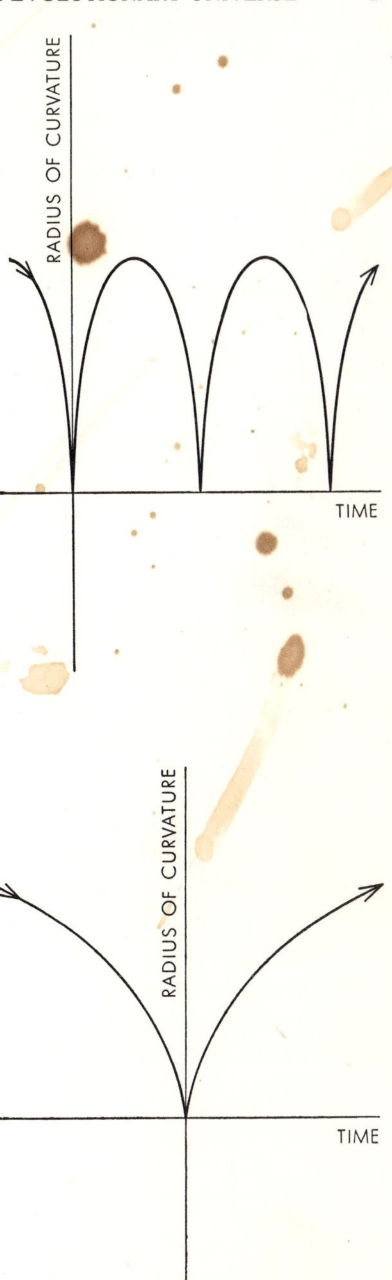

PULSATING AND HYPERBOLIC universes are represented by curves. The pulsating universe at the top repeatedly expands to a maximum permissible density and contracts to a minimum permissible density. The hyperbolic universe at the bottom contracts and then expands indefinitely.

We have reviewed the questions that dominated the thinking of cosmologists during the first half of this century: the conception of a four-dimensional space-time continuum, of curved space, of an expanding universe and of a cosmos which is either finite or infinite. Now we must consider the major present issue in cosmology: Is the universe in truth evolving, or is it in a steady state of equilibrium which has always existed and will go on through eternity? Most cosmologists take the evolutionary view. But in 1951 a group at the University of Cambridge, whose chief spokesman has been Fred Hoyle, advanced the steady-state idea. Essentially their theory is that the universe is infinite in space and time, that it has neither a beginning nor an end, that the density of its matter remains constant, that new matter is steadily being created in space at a rate which exactly compensates for the thinning of matter by expansion, that as a consequence new galaxies are continually being born, and that the galaxies of the universe therefore range in age from mere youngsters to veterans of 5, 10, 20 and more billions of years. In my opinion this theory must be considered very questionable because of the simple fact (apart from other reasons) that the galaxies in our neighborhood all seem to be of the same age as our own Milky Way. But the issue is many-sided and fundamental, and can be settled only by extended study of the universe as far as we can observe it. Hoyle presents the steady-state view in "The Steady-State Universe" [September 1956]. Here I shall summarize the evolutionary theory.

We assume that the universe started from a very dense state of matter. In the early stages of its expansion, radiant energy was dominant over the mass of matter. We can measure energy and matter on a common scale by means of the well-known equation $E=mc^2$, which says that the energy equivalent of matter is the mass of the matter multiplied by the square of the velocity of light. Energy can be translated into mass, conversely, by dividing the energy quantity by c^2. Thus we can speak of the "mass density" of energy. Now at the beginning the mass density of the radiant energy was incomparably greater than the density of the matter in the universe. But in an expanding system the density of radiant energy decreases faster than does the density of matter. The former thins out as the fourth power of the distance of expansion: as the radius of the system doubles, the density of radiant energy drops to one sixteenth. The density of matter declines as the third power; a doubling of the radius means an eightfold increase in volume, or eightfold decrease in density.

Assuming that the universe at the beginning was under absolute rule by radiant energy, we can calculate that the temperature of the universe was 250 million degrees when it was one hour old, dropped to 6,000 degrees (the present temperature of our sun's surface) when it was 200,000 years old and had fallen to about 100 degrees below the freezing point of water when the universe reached its 250-millionth birthday.

This particular birthday was a crucial one in the life of the universe. It was the point at which the density of ordinary matter became greater than the mass density of radiant energy, because of the more rapid fall of the latter [*see chart on this page*]. The switch from the reign of radiation to the reign of matter profoundly changed matter's behavior. During the eons of its subjugation to the will of radiant energy (*i.e.*, light), it must have been spread uniformly through space in the form of thin gas. But as soon as matter became gravitationally more important than the radiant energy, it began to acquire a more interesting character. James Jeans, in his classic studies of the physics of such a situation, proved half a century ago that a gravitating gas filling a very large volume is bound to break up into individual "gas balls," the size of which is determined by the density and the temperature of the gas. Thus in the year 250,000,000 A. B. E. (after the beginning of expansion), when matter was freed from the dictatorship of radiant energy, the gas broke up into giant gas clouds, slowly drifting apart as the universe continued to expand. Applying Jeans's mathematical formula for the process to the gas filling the universe at that time, I have found that these primordial balls of gas would have had just about the mass that the galaxies of stars possess today. They were then only "protogalaxies"—cold, dark and chaotic. But their gas soon condensed into stars and formed the galaxies as we see them now.

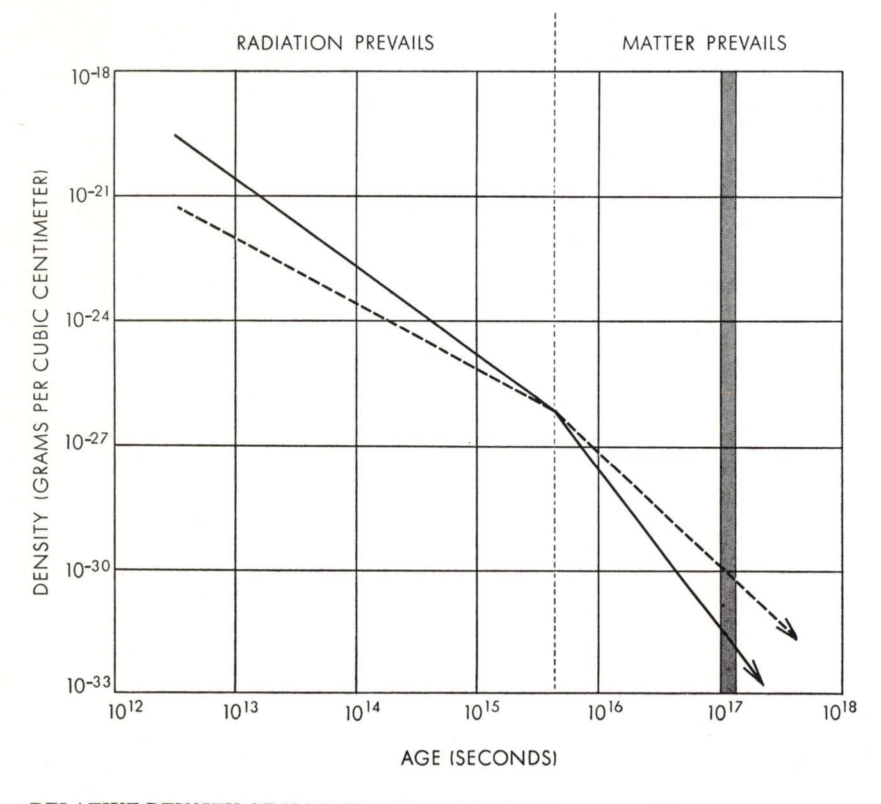

RELATIVE DENSITY OF MATTER AND RADIATION is reversed during the history of an evolutionary universe. Up to 250 million years (*broken vertical line*) the mass density of radiation (*solid curve*) is greater than that of matter (*broken curve*). After that the density of matter is greater, permitting the formation of huge gas clouds. The gray line is the present.

A central question in this picture of the evolutionary universe is the problem of accounting for the formation of the varied kinds of matter composing it—*i.e.*, the chemical elements. The question is discussed in detail in "The Origin of the Elements" [September 1956]. My belief is that at the start matter was composed simply of protons, neutrons and electrons. After five minutes the universe must have cooled enough to permit the aggregation of protons and neutrons into larger units, from deuterons (one neutron and one proton) up to the heaviest elements. This process must have ended after about 30 minutes, for by that time the temperature of the expanding universe must have dropped below the threshold of thermonuclear reactions among light elements, and the neutrons must have been used up in element-building or been converted to protons.

To many a reader the statement that the present chemical constitution of our universe was decided in half an hour five billion years ago will sound nonsensical. But consider a spot of ground on the atomic proving ground in Nevada where an atomic bomb was exploded three years ago. Within one microsecond the nuclear reactions generated by the bomb produced a variety of fission products. Today, 100 million million microseconds later, the site is still "hot" with the surviving fission products. The ratio of one microsecond to three years is the same as the ratio of half an hour to five billion years! If we can accept a time ratio of this order in the one case, why not in the other?

The late Enrico Fermi and Anthony L. Turkevich at the Institute for Nuclear Studies of the University of Chicago undertook a detailed study of thermonuclear reactions such as must have taken place during the first half hour of the universe's expansion. They concluded that the reactions would have produced about equal amounts of hydrogen and helium, making up 99 per cent of the total material, and about 1 per cent of deuterium. We know that hydrogen and helium do in fact make up about 99 per cent of the matter of the universe. This leaves us with the problem of building the heavier elements. I hold to the opinion that some of them were built by capture of neutrons. However, since the absence of any stable nucleus of atomic weight 5 makes it improbable that the heavier elements could have been produced in the first half hour in the abundances now observed, I would agree that the lion's share of the heavy elements may well have been formed later in the hot interiors of stars.

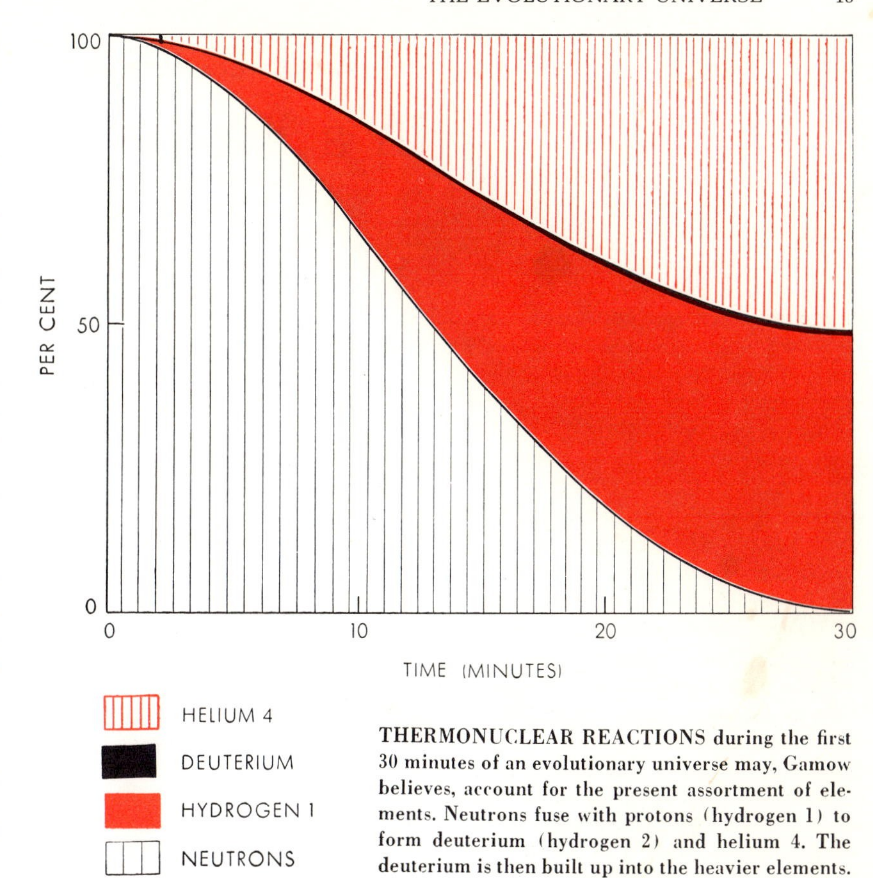

THERMONUCLEAR REACTIONS during the first 30 minutes of an evolutionary universe may, Gamow believes, account for the present assortment of elements. Neutrons fuse with protons (hydrogen 1) to form deuterium (hydrogen 2) and helium 4. The deuterium is then built up into the heavier elements.

All the theories—of the origin, age, extent, composition and nature of the universe—are becoming more and more subject to test by new instruments and new techniques, which are described in later articles in this issue. In the article on the red-shift investigations, Allan Sandage reports a tentative finding that the expansion of the universe may be slowing down. If this is confirmed, it may indicate that we live in a pulsating universe. But we must not forget that the estimate of distances of the galaxies is still founded on the debatable assumption that the brightness of galaxies does not change with time. If galaxies actually diminish in brightness as they age, the calculations cannot be depended upon. Thus the question whether evolution is or is not taking place in the galaxies is of crucial importance at the present stage of our outlook on the universe.

3

The Curvature of Space in a Finite Universe

by J. J. Callahan
August 1976

Curvature of a surface is an intrinsic property that gives rise to distortion of distances on a map. The same is true for curvature of space, where the map is Einstein's general theory of relativity

Is the universe finite or infinite? According to numerous ancient mythologies, it has a complex structure but is nonetheless finite. That viewpoint developed in Greek philosophy and culminated in the cosmology of Eudoxus and Aristotle: the earth is a ball surrounded by a series of concentric crystalline spheres, the outermost sphere carrying the fixed stars and containing within it the entire material universe. The primary purpose of this cosmology was to explain the motions of the planets and other celestial bodies. Each body was carried around the earth by the rotation of the sphere in which it was embedded. Nevertheless, an integral part of the theory was that the universe was finite.

Aristotle's picture of the world was widely accepted in medieval Europe; it appears, for example, in scholastic philosophy and in Dante Alighieri's *Divine Comedy*. In fact, Dante actually extended Aristotle's picture in a radical and thoroughly modern way. I shall take up Dante's interpretation in my conclusion. In spite of the popularity of the finite-world picture, however, it is open to a devastating objection. In being finite the world must have a limiting boundary, such as Aristotle's outermost sphere. That is impossible, because a boundary can only separate one part of space from another. This objection was put forward by the Greeks, reappeared in the scientific skepticism of the early Renaissance and probably occurs to any schoolchild who thinks about it today. If one accepts the objection, one must conclude that the universe is infinite.

The notion of infinity has always been wrapped in mystery, and historically it triggered apprehensions that have only gradually been overcome. During the scientific Renaissance, Euclidean geometry became the main instrument for comprehending infinite physical space. Euclidean geometry contends that a straight line is the shortest distance between two points, and that the sum of the angles in a triangle will always be 180 degrees. The Renaissance scientists saw that Euclidean geometry treated ideal objects in a mathematical context that was infinite, but that its axioms and propositions exactly described the spatial relations of the real world. Leibniz and Newton shared the view that physical space was infinite and Euclidean. They disagreed, however, on how matter was situated in space. For Leibniz a finite group of stars was unthinkable: such a group would have to be in some specific location in space and God would have had no sufficient reason to put it in one place rather than in some other. Leibniz thus concluded that the universe must be infinite. Newton rejected that possibility, however, on the grounds that God is the only possible actual infinity. Although today these arguments may not seem persuasive, at the time they were considered sufficient.

Who was right? Both arguments were essentially negative. Leibniz denied that the universe was finite, Newton denied that it was infinite. Neither was enthusiastic, however, about the alternative with which he was left. In 1781 Immanuel Kant offered in his *Critique of Pure Reason* a thorough analysis of the entire problem of space, including a bold and novel resolution of the dispute between Newton and Leibniz. Kant said that they were both right and that we must admit paradoxically that the universe is neither finite nor infinite! This basic contradiction between principles that seem equally necessary and reasonable is known as Kant's antinomy of space. The antinomy of space is one of several antinomies that in Kant's view pointed to "a hereditary fault in metaphysics that cannot be explained, much less removed, except by ascending to its birthplace, pure reason itself." A major aim of the *Critique of Pure Reason* was to remove such hereditary faults from metaphysics. Kant's method was drastic. He argued that since we cannot conceive of the universe as being either finite or infinite, we shall never be able to discover empirically whether it is either finite or infinite. In other words, it is not an objective property of the universe to be either finite or infinite. Furthermore, space is not a thing but is a form through which we perceive things, and we make a fundamental error when we treat space as a thing. The antinomy reflected a basic limitation in the mental processes we use to describe the world. Kant would insist that we discard our question as being meaningless.

Today Kant's metaphysical analysis of space is disregarded by modern science because its foundation—notably Euclidean geometry—has been broadened by revolutionary developments in mathematics and physics. Einstein's general theory of relativity provides a new geometry of space, and it opens another approach, unforeseen by Kant, to the question about the finiteness of the world. For Kant the question had simply been invalid. Einstein restored its validity by arguing that Kant's antinomy of space is only apparent and that it can be understood without resorting to metaphysics. In short, Einstein shows that a finite universe is a real possibility.

Like any other physical theory, the general theory of relativity deals with matter and its properties. It regards a galaxy as being perhaps the most natural unit of matter on the cosmic scale. Thus at this level the problem of space is the problem of understanding how the galaxies fit together. A convenient way of visualizing Einstein's solution to the problem is to construct a laboratory model of the entire galactic system. One could construct the model out of balls and sticks, like a model of a large molecule, except that each ball would represent a galaxy and each stick the distance between two galaxies. Before examining Einstein's model, however, let me go back and translate the views of Newton, Leibniz and Kant into the language of models. That may demonstrate more clearly both what they said and how the general theory of relativity went beyond them.

Let us turn first to Newton and his assertion that the universe is finite. If Newton is right, then there can be only a finite number of galaxies. Furthermore, it is reasonable to expect that in time one might devise instruments for locating all of them, and then a complete model of the galactic system could be built. In any case such a model is possible as a mental construct, and that is sufficient for our purposes. If Newton had thought in terms of such a model, he certainly would have had in mind an exact scale model of the universe: one in which the distances between balls were exactly proportional to the distances between the galaxies they represented.

The main consequence of exact scaling is that any metric feature of the model (that is, any feature that depends only on distance) will be shared by the galactic system. In other words, the model and the galactic system should have the same metric features. The laws of Euclidean geometry are known through direct observation to hold in the terrestrial laboratory, and they dictate all the metric properties of the model. Hence those same laws must dictate the metric properties of the galactic system. That conclusion is very important; in fact, it is the key to everything that follows. It states that the laws of Euclidean geometry are valid in the galactic system not because they are directly verifiable through observation and measurement in the galactic system but because the system can be reproduced in a scale model. The converse will also be true: if it is impossible to reproduce the galactic system in an exact scale model, then we must abandon the conviction that the geometry of intergalactic space is Euclidean.

Now, a geometric figure and any scale model of it are similar, meaning that corresponding angles in the figure and in the model are identical and corresponding sides are directly proportional to each other in length. Thus we are essentially saying that space can contain simi-

SPACE WAS FINITE and had a definite edge, according to the Aristotelian cosmology accepted during medieval times. Here a man is shown looking beyond the edge of space to the Empyrean abode of God beyond. The illustration is often said to be a 16th-century German woodcut; according to Owen Gingerich of Harvard University, it is more likely a piece of art nouveau that was apparently published for the first time in 1907 in *Weltall und Menschheit,* edited by Hans Kraemer. In either case the picture clearly demonstrates a dilemma posed by Immanuel Kant known as Kant's antinomy of space. Kant believed that the universe had to be finite in extent and homogeneous in composition, and that space had to obey the laws of Euclidean geometry. Actually, however, all those assumptions cannot be true at once. Newton, Leibniz and Einstein had different ways of resolving the dilemma, shown in illustrations on next two pages.

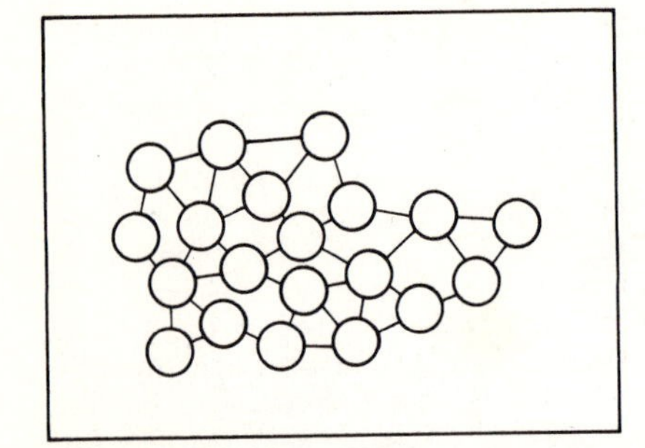

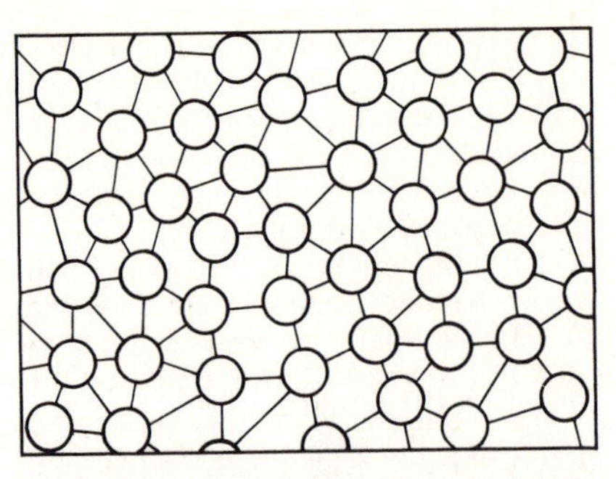

TWO "BALL AND STICK" COSMOLOGICAL MODELS demonstrate the philosophical views of Newton and Leibniz; each ball represents a galaxy and each stick represents the distance between galaxies. Although both men accepted Kant's assumption that space obeyed Euclidean geometry, Newton believed the galactic system was finite and inhomogeneous (*left*); thus his model of the galactic system had both a center and a boundary. Leibniz, however, believed the galactic system was infinite and homogeneous (*right*), with no center or boundary.

lar figures of arbitrary sizes if and only if its structure is given by Euclidean geometry. That result is actually very old: it was first obtained by the English mathematician John Wallis in 1663. In either case, using the language of models or of similar figures, we end up with a criterion for determining the geometric properties of space.

Any model of a finite universe has two metric features of special interest. First, the model has a "geographic" center. Second, it has a boundary, consisting of those balls with neighbors on only one side. Therefore if Newton is right, the galactic system must also have a center and a boundary, because it possesses all the metric properties of its scale model.

It is the inhomogeneity of Newton's universe (the fact that not all galaxies have neighbors on all sides) and not its finiteness as such that Leibniz could not accept. Any finite model has a boundary and a center; in order to eliminate those features one would have to add an infinite number of new balls. Then the model becomes impossible to build. Nevertheless, it is possible to imagine an arrangement of balls reproducing exactly the arrangement of galaxies and fading into the distance in all directions. Let us take that mental construct as Leibniz' scale model. It resembles a model one can build of a crystal, which is also infinite in a theoretical sense. Hence Leibniz' model is actually no stranger than Newton's. If the galaxies are more or less uniformly distributed, then the model will have neither a center nor a boundary.

Like Leibniz, Newton and everyone else in the 18th century, Kant believed in the validity of Euclidean geometry. Unlike many, he knew that Euclidean geometry could not be justified by experience alone. In fact, our key argument, that we judge space to be Euclidean not through observation but by constructing a model of the galactic system, is due to Kant. In the *Critique of Pure Reason* he actually did not address either models or galaxies. He declared that a direct intuition of space—what I am calling a model—is given to each of us, and through it we discover the properties of space. Because the intuition is universally shared among human beings, it is one of the dictates of pure reason and must be put on an equal footing with sensory experience in investigations of the world. What is more, since the intuition does not depend on experience, which can be faulty or incomplete, the knowledge intuition gives must necessarily be true. As Kant put it, Euclidean geometry is synthetic a priori, by which he meant that it is a special kind of knowledge that is truly descriptive of the world of experience but is not itself derived from that experience. Just as today every science strives to become exact, in the 18th century classical physics saw geometry as its ideal. Kant made his beliefs explicit because he wanted to exploit the special status of geometry to refute the claim of empiricists, most successfully advanced by David Hume, that *all* knowledge of the world is sensory. But is such a direct intuition absolutely necessary to understanding the universe? Must the world admit a scale model? Something fundamental in Kant's philosophy would be undermined if a different situation were to be perceived.

Newton's and Leibniz' models are two clearly distinct models of the galactic system. Kant rejected them both. He had to, because each lacked what he felt was an essential property. For his part he maintained that any study of the material universe must begin by acknowledging three facts: first, the galactic system is finite; second, it is homogeneous and unbounded; third, it can be reproduced in an exact scale model. Those three facts, however, cannot be simultaneously true. In other words, no exact scale model can be both finite and homogeneous. Once again we have arrived at Kant's antinomy of space, this time through the language of models.

Why does the antinomy arise? Kant blamed it on inherent limitations in the mental processes we use to describe the world. In this case the mental process is model building. Kant is saying that we are unable to build a model of the galactic system because our minds cannot tell us how.

There is a way to get out of the antinomy: simply refuse to accept one of Kant's "facts." Remember, not one of them is a physical fact established by direct observation. They are all intuitive assumptions. Even Kant admitted that.

EINSTEIN'S COSMOLOGICAL MODEL achieves Kant's desideratum of a finite and homogeneous universe, but only by rejecting Kant's assumption that space was Euclidean. To say that space at large is curved means that it may not be possible to build an exact scale model of the galactic system in the laboratory, where the laws of Euclidean geometry rule. For example, five equidistant galaxies may exist in space, but any attempt to construct a scale model of such a system will fail (*left*). Einstein resolves Kant's antinomy of space by suggesting that when one builds the ball-and-stick model of the system, one should just say that one long stick joining two of the galaxies represents same distance in space as the nine shorter sticks (*right*).

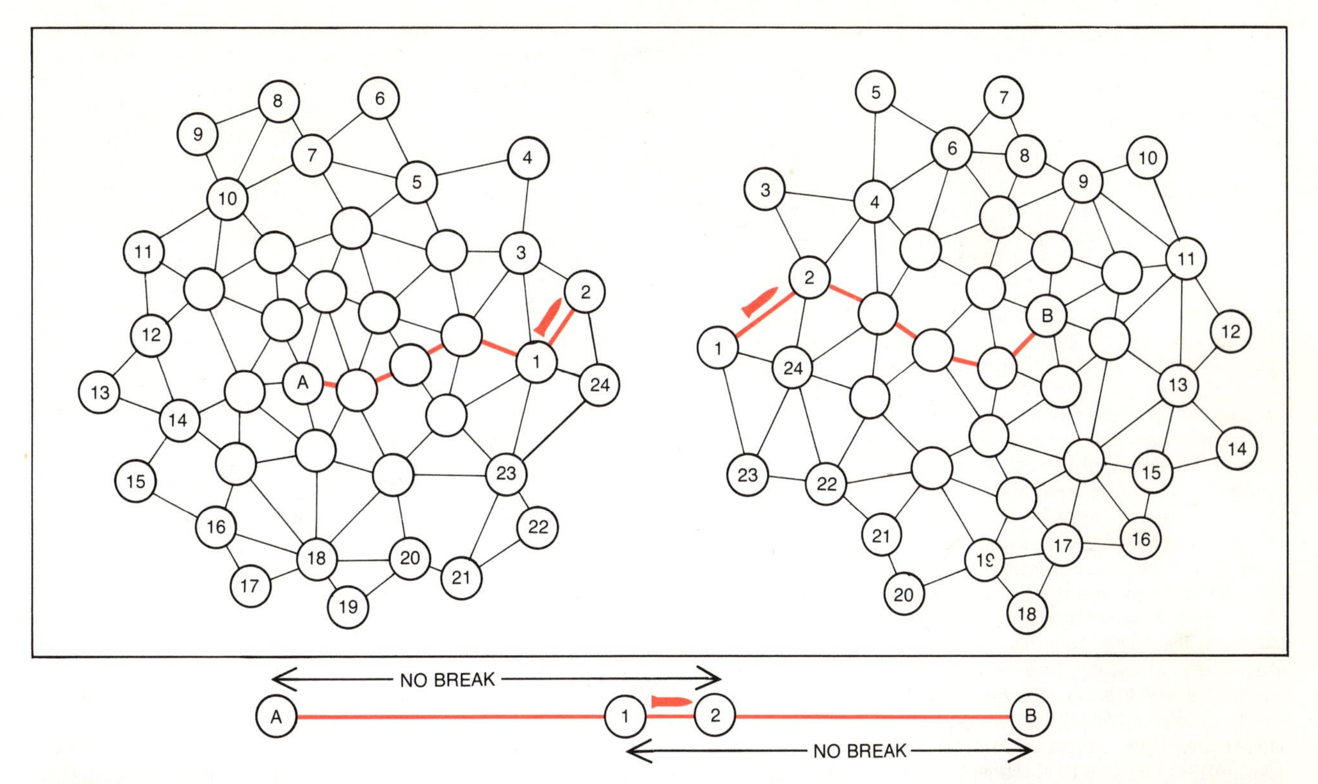

EINSTEIN'S MODEL for a finite galactic system is shown here in somewhat more detail. The model is entirely contained within the limits of the illustration. The model appears to be highly inhomogeneous: it is shown in two disconnected pieces each with a center and a boundary, and not all galaxies appear to have neighbors on all sides. Furthermore, two numbered balls, one in each piece, represent the same galaxy. Unnumbered balls represent distinct galaxies that are not duplicated. The model is not built to scale, however; therefore its peculiar properties are not necessarily reflected in the galactic system itself. In fact, the galactic system is actually quite homogeneous. The color line between ball *A* and ball *B* represents a continuous path through the galactic system being taken by a rocket (also shown twice). The continuous nature of the rocket's path is illustrated at bottom. One can think of the model as a pair of three-dimensional viewing screens in which every galaxy appears on at least one screen, and any galaxy at the edge of one screen also appears on the other.

One possibility is to follow Newton and reject Kant's second assumption that the galactic system is homogeneous and unbounded. Then there is no problem in accepting the remaining two assumptions and building a suitable model. That may in fact be the correct choice, since a finite but inhomogeneous galactic system has never been ruled out experimentally. In any case the antinomy is gone. Another possibility is to follow Leibniz and reject the first assumption, that the galactic system is finite. This choice also removes the antinomy, since a homogeneous but infinite system has not been ruled out experimentally either. Or, finally, we can follow Einstein.

Einstein actually achieved Kant's ambition by constructing a down-to-earth model of a finite homogeneous galactic system. He did it by rejecting Kant's third assumption, that the model must be exactly scaled. For example, in Einstein's model a three-inch stick in one location may represent an intergalactic distance of, say, 50 million light-years, whereas in another location it may represent 60 million light-years. To see the impact of Einstein's innovation consider the following imaginary scene.

One day at some time in the distant future the members of an intergalactic surveying team return to their home base, having measured all the distances between five galaxies in which they are particularly interested. They exchange information and discover that each of the five galaxies is equidistant from the other four. Immediately they can stick four balls together at the vertexes of a regular tetrahedron to represent four of the equidistant galaxies. Where should the fifth ball go? They can attach it to any three of the other four balls and form a second tetrahedron. It is then equidistant from the three balls but not from the fourth ball. Actually there is no way to build a laboratory model consisting of five equidistant balls. Forget about a grand design for the entire galactic system; here is a mere handful of galaxies presenting a crisis in intuition. As a spokesman for the Euclidean position, Kant would have argued that the surveyors are mistaken, because space "is not like that." But the surveyors have already rechecked their work and there is no mistake. Space *is* like that. The assumption that we can build a scale model of any physical system, which is equivalent to the assumption that the geometry of space is Euclidean, is thus revealed to be an attempt to make reality conform to our preconceptions. Einstein turns that around and makes the model conform to reality: he takes the model already built and declares that one long stick should represent the same intergalactic distance as the other nine joining all the balls.

It is only a short step from five galaxies to Einstein's model of the entire galactic system. The illustration above presents a simplified version of Einstein's model using several dozen balls instead of the millions that might actually be required. The number of balls does not really matter, since a larger model will exhibit the same essential features. Two features of the model are particularly noticeable. First, in this illustration there are two disconnected pieces, each with a boundary; second, sometimes two different balls, one in each piece, represent the same galaxy. Because the model is not built exactly to scale, however, those peculiar properties need not carry over to the galactic system itself. In fact, the system is actually quite homogeneous, as one can see by thinking of the model as a pair of three-dimensional viewing screens. In that case every galaxy appears in at least one of the screens, and to guarantee that none is overlooked, any galaxy appearing at the edge of one screen also appears at the edge of the other screen. That explains why two balls sometimes represent the same galaxy.

Now look at one of the galaxies that is visible in both screens. Half of its neighbors appear in one screen and half in the other. Thus although each ball representing that one galaxy is on the boundary of its own screen, the galaxy itself must be completely surrounded by neighboring galaxies. Since the remaining galaxies appear in the middle of either one screen or the other, they also have neighbors on all sides. Hence the galactic system has no boundary. Furthermore, the galactic system is connected, because an object can move continuously from any one galaxy to any other, although it may be necessary to switch viewing screens at some point in order to follow the motion.

Distance distortion is a third essential feature of the model, but it is not particularly noticeable in such an incomplete illustration. Exactly how much distances are distorted is a technical matter, fully treated in Einstein's detailed model (the general theory of relativity). The simple presence of the distortion is what concerns us, because it reveals a fundamental property of space, that is, of the metric relations of the galactic system. Whether or not a model must distort distances is determined solely by the nature of space. As we have already seen, if the galactic system admits an exact scale model, space is Euclidean. If no scale model is possible, that fact must likewise be due to some contrary property of space. This property has been given the name curvature.

The word curvature may seem an odd choice for a property of space, and so I shall discuss the reason for it below. In any event curvature of space refers simply to the need for distorted models and has nothing to do with space being somehow mysteriously bent. Taking Einstein's model into consideration, one can now draw one of the fundamental conclusions of the general theory of relativity: Kant's antinomy of space arises from an unwarranted assumption that space is Euclidean. A finite and homogeneous galactic system is conceivable, and it is the curvature of space that makes it so.

The connection between curvature and a finite world is a common subject in popular accounts of relativity, but it is usually explained by drawing an analogy with what can happen on surfaces. On a flat plane any finite set of points has a boundary. On the spherical surface of the earth, however, one can imagine a more or less uniformly distributed network of, say, weather stations, and although there are a finite number of stations, every station has neighbors on all sides, so that the network has no boundary. The analogy suggests that the space of a finite homogeneous galactic system must therefore be like the surface of the earth, only one dimension larger. Now, the earth's surface is only two-dimensional, which means that the position of a point on it is determined by just two numbers, latitude and longitude. The reason the earth's surface has no boundary, we feel, is that it curves around into a third dimension. We are led to infer that a three-dimensional "spherical" space must somehow curve around into a fourth dimension. The analogy collapses because it is hopeless to imagine what the extra spatial dimension looks like; no one has ever seen it.

The analogy is actually a good one, but oddly enough it suffers from too detailed a picture of the earth and not too scant a picture of space. A surface has two kinds of geometric properties, intrinsic and extrinsic. An intrinsic property is one that refers to measurements that can be carried out entirely on the surface itself; any other property is extrinsic. For example, in the third century B.C. Eratosthenes deduced the radius of the earth by combining an intrinsic fact about the earth's surface (the fact that the distance from Syene to Alexandria is 5,000 stadia) with an extrinsic fact (the fact that when the sun is directly overhead at Syene, it is 7.2 degrees from the zenith at Alexandria).

Physical space has no extrinsic geometry that we know of, however, because every spatial property we know relates to figures and measurements made in space itself. Thus space cannot be analogous to the surface of a sphere, because it has nothing comparable to the sphere's extrinsic geometry. An analogy between different objects can exist in our minds only where we find it possible to overlook their differences.

Still, there is a way to conceive of the curvature of space. Our actual goal is to understand what a finite homogeneous galactic system must be like. Here it is useful to consider the example of a network of weather stations on the surface of the earth. Notice that the relevant geometric properties of the network are all intrinsic: each weather station is surrounded by neighbors, and the network of stations, although finite, has no boundary. Since those are intrinsic properties, we lose nothing by ignoring the earth's extrinsic geometry; moreover, since the extrinsic part was the stumbling block all along, we are even better off ignoring it. Irrelevant information can be just as distracting to a mathematician as it is to a reader of mystery stories. In mathematics the job of isolating what is genuinely relevant is called abstraction.

The language of models can help once again. A terrestrial globe is an exact

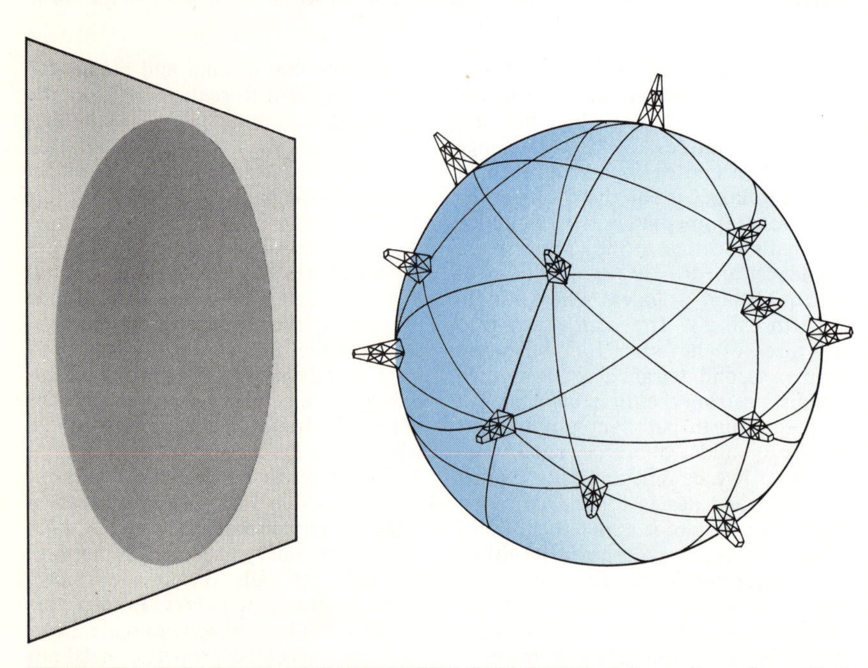

INTRINSIC GEOMETRY AND EXTRINSIC GEOMETRY of the surface of an object such as a sphere are quite different. Any intrinsic property of a surface refers to measurements that can be carried out on the surface itself; any other property is extrinsic. For example, the fact that on a sphere a network of weather stations can be both finite and unbounded is an intrinsic property. The fact that the sphere casts a circular shadow from all viewpoints is an extrinsic geometric feature. Physical space has no known extrinsic geometry because every property we know of relates to figures and measurements made in space itself. Thus it is hopeless to try to imagine curved space as being mysteriously bent through a fourth dimension, since we cannot take an extrinsic view of space by getting outside it and looking back at it.

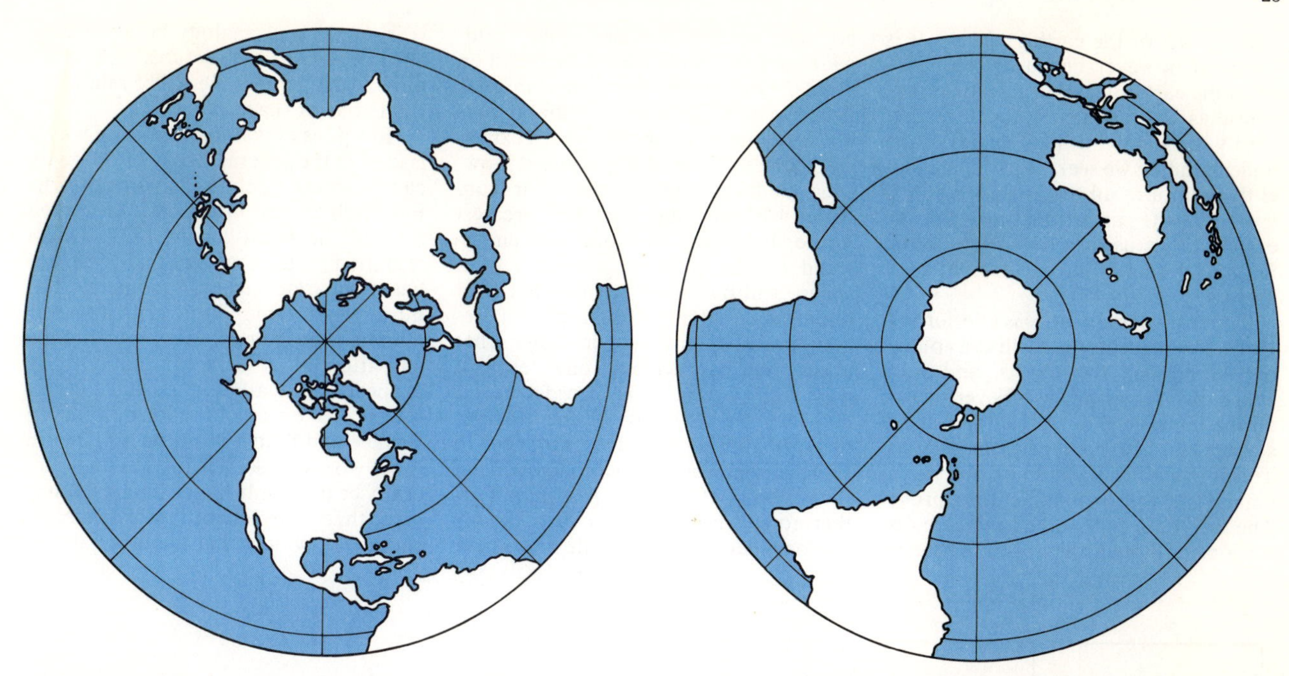

ATLAS OF CHARTS of the spherical earth captures all the intrinsic properties of the earth while filtering out the extrinsic geometry. Since the curved surface of the earth has been flattened out on the charts, distances are distorted. The amount and kind of distortion contain all the information needed to reconstruct the full geometry of the earth. Hence we no longer need a third dimension in order to understand the curvature of the surface of a sphere; similarly, we do not need a fourth dimension to understand the curvature of space.

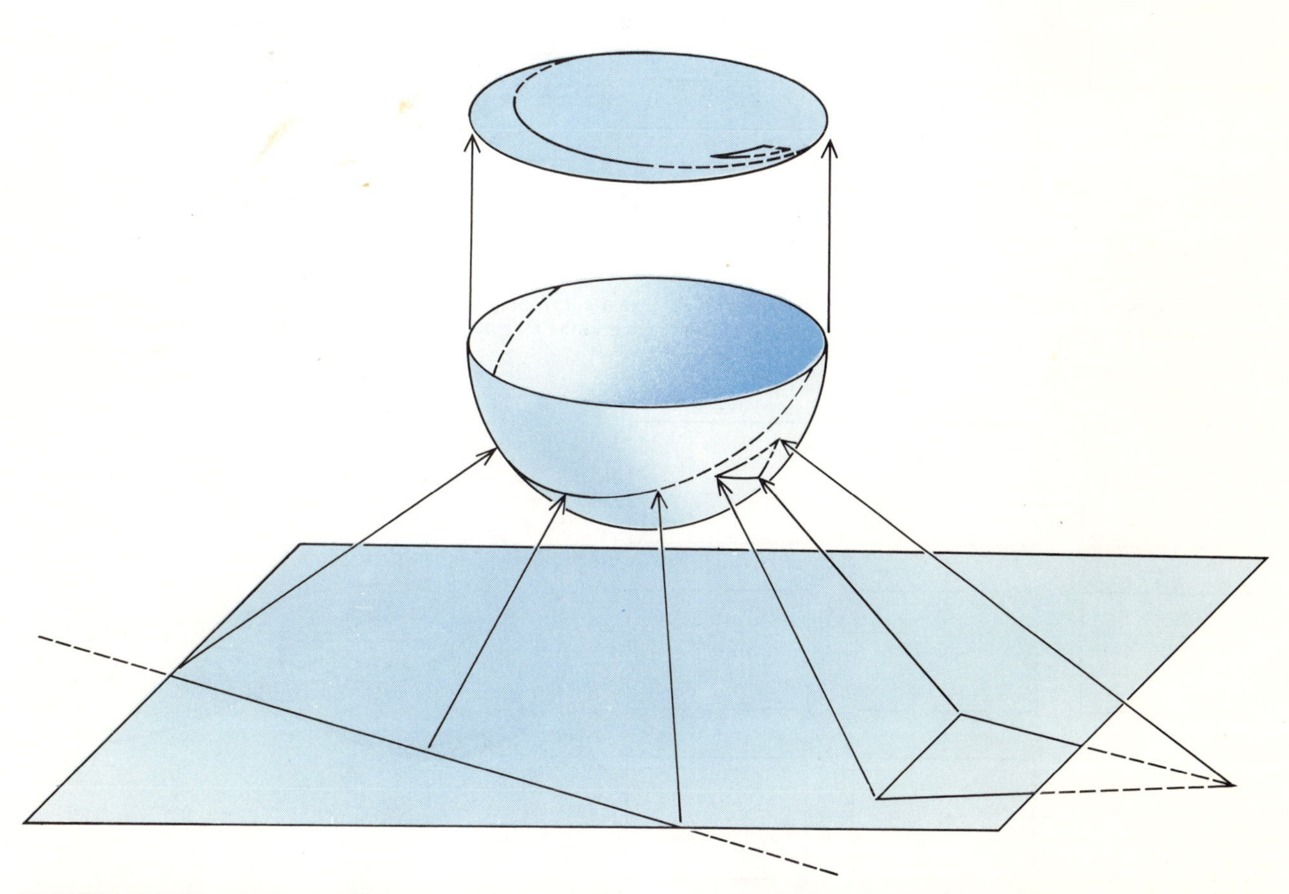

INFINITE PLANE CAN BE MAPPED onto an atlas of charts as well as a sphere can. In this particular case the atlas consists of one chart, the disk, which is made by first projecting the plane onto the hemisphere and then up onto the disk. (This is only one of many ways of mapping the plane.) The geometric shapes again reveal the distance distortions created by the mapping. Any unbounded two-dimensional surface can be mapped by an atlas of charts; such a surface, including the plane and sphere, is a two-dimensional manifold.

scale model of the earth's surface. It is too good, however, because it reproduces all the earth's geometric features, extrinsic as well as intrinsic, with perfect clarity. Ironically, using it denies us the understanding we seek. We need a model that captures all the intrinsic geometry but at the same time filters out the extrinsic. Let us turn now to the other familiar model of the earth: an atlas of maps.

In many ways any atlas is inferior to a globe. It represents the earth as a collage of overlapping flat charts, and each chart distorts distances. Nevertheless, all the intrinsic geometry of the earth's surface can be recovered from an atlas. This is actually a surprising result mathematically, and it is difficult to prove. The fact remains that the inevitable distortions, although they are a nuisance, are manageable. How else could worldwide air and sea navigation be based on charts? One might object that an atlas does not look like the earth, that it does not help one to grasp the earth's extrinsic geometry. For our purposes, however, the flatness of the charts, far from being a disadvantage, is their greatest virtue. It implies that one can understand all the intrinsic geometry of a sphere without ever leaving a flat two-dimensional plane. That fact is a tremendous economy, and it shows what abstraction can achieve: one does not need a third dimension in order to understand the structure of the weather network; hence one does not need a fourth dimension in order to understand the structure of the galactic system. Furthermore, what seems to be a creaky analogy between the earth and space is in fact a real analogy between their abstract intrinsic structures: Einstein's model (the general theory of relativity) is an atlas of space.

How does curvature fit in? The way it is used in the general theory of relativity can be traced back to the work of Carl Friedrich Gauss, in a theory of curved surfaces he published in 1827. Gauss was the first to recognize that a surface has a separate intrinsic geometry. His most remarkable discovery, however (and he even called it that), was that the curvature of a surface is an intrinsic property. Basically Gauss said the following: Take a small portion of a curved surface and flatten it out on a plane. In other words, make a chart. This can generally be done only by stretching the surface, that is, by distorting distances. It should be evident that the original cur-

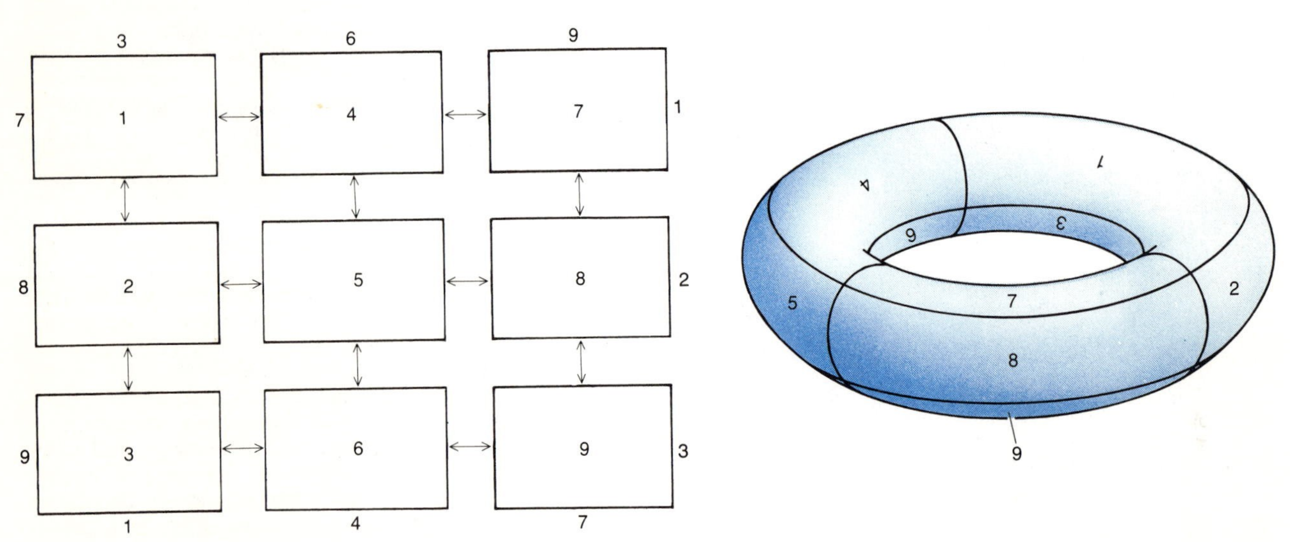

AN ATLAS FOR A TORUS (*left*) might consist of nine charts; the numbers around the edges of each chart indicate which of the other charts it overlaps. The numbered regions on the torus corresponding to the nine different charts in the atlas are shown at the right.

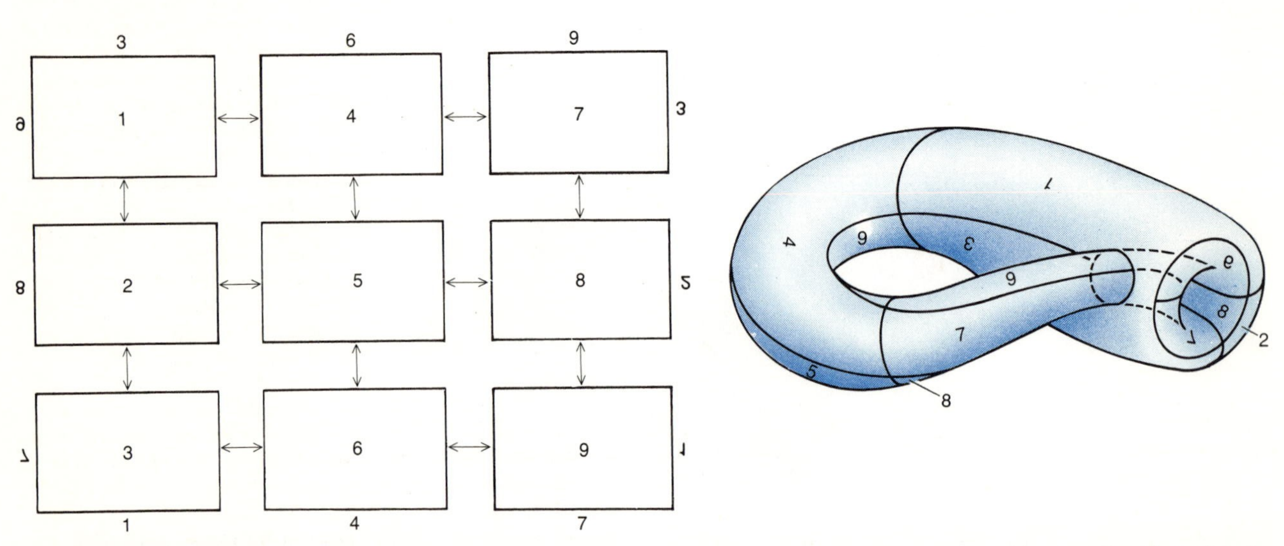

AN ATLAS FOR A KLEIN BOTTLE (*left*) might also consist of nine charts. It is very similar to the atlas for the torus except in the way in which the charts overlap. That change, however, makes it impossible to paste the charts together without making the surface intersect itself in places where it should not. The Klein bottle is a form that has one surface. It has no "inside," as a torus and a sphere do.

vature of the surface determines the precise kind and amount of distortion. Gauss's remarkable discovery is the converse of that fact: the curvature of a surface can be completely determined solely from the distance distortions present in a chart. In other words, distortion of a chart and curvature of a surface are different aspects of the same thing: when the intrinsic geometry of a portion of a surface is abstracted to a chart, its curvature is given by distance distortion. That is why the completely analogous distortions in Einstein's model of the galactic system are attributed to the curvature of space.

Gauss studied the intrinsic geometry of a small portion of a surface by making a chart. In order to map an entire surface it is generally necessary to resort to several charts, or an atlas. A surface that can be mapped by an atlas of charts is called a two-dimensional manifold. The term is meant to emphasize that the surface is usually made up of many two-dimensional patches instead of a single chart.

Every smooth surface without a boundary is a two-dimensional manifold. For example, an infinite plane is a two-dimensional manifold, and so is a torus. One can also start with any arbitrary collection of charts that overlap in a coherent way; they constitute an atlas for some two-dimensional manifold.

The notion of a manifold, which grew out of the attempt to understand the intrinsic geometry of a surface, actually yields something more general. One can make charts out of solid, three-dimensional "blobs" instead of flat, two-dimensional "patches" and have something that makes sense in three dimensions. The resulting object is called, naturally enough, a three-dimensional manifold, and it is a volume rather than a surface. Einstein's model of the universe is an atlas for one particular three-dimensional manifold: it is called the three-dimensional sphere, or three-sphere. (The surface of an ordinary sphere is a two-sphere.) There are countless other examples. The ordinary Euclidean space of mathematics is given by a single chart, analogous to the chart for the infinite plane. Other spaces quickly become too messy to describe in detail. Mathematicians have even defined manifolds for arbitrary numbers of dimensions, by finding a way of getting around the need to visualize the fundamental building blocks: the dimension of a manifold is simply interpreted as the number of variables needed to locate a point on it. Manifolds have thus become a natural setting for problems requiring many variables, and their use now extends beyond mathematics to science in general.

One of the most basic perceptions we have of our environment is that three variables (height, width and breadth, or *x*, *y* and *z*) are needed to label completely all positions within it. Physical space is a three-dimensional manifold. The mistake of common sense was to assume that it must be the one given by a single chart: Euclidean space. We can now see why this was such a natural error to fall into. Locally (that is, in the immediate neighborhood of any point) all manifolds of the same dimension look alike. In principle they are distinguishable because of the presence of curvature; in practice, however, curvature may not be detectable with sufficient experimental accuracy when it is measured over a small region. (The surveyors in my little scene had to travel to other galaxies to get their results; our own measuring instruments have not yet left the solar system.) In other words, for all practical purposes space *is* Euclidean if it is taken in small enough pieces. The entire cosmos is another matter, and the notion of the manifold provides a wealth of new possibilities for understanding the structure of the universe in the large. Far from being bankrupted by the apparent nonsense of questions about the structure of space, as Kant suggested, mathematics has been substantially enriched.

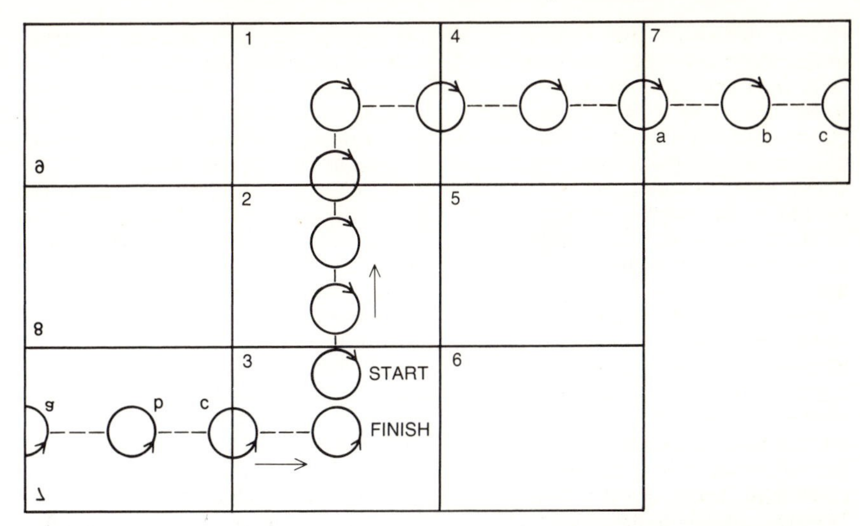

ORIENTABILITY is a global property of a surface. A circle with an arrowhead on its circumference is said to define a local orientation (either clockwise or counterclockwise) when it is drawn on a chart for a manifold. That orientation can be extended to other charts by moving the circle from one chart to another in the atlas. Moving along two different routes, however, may produce conflicting orientations of the circular arrowhead in a distant chart. If that happens, the manifold is said to be globally nonorientable. A Klein bottle is globally nonorientable; shown is its atlas with one chart (*chart 7*) and three of the positions of the circular arrowhead (*a–c*) included twice. When the circular arrowhead returns to its starting position in chart 3, it is an upside-down mirror image of its former self. In some manifolds, for example the torus, such a conflict can never arise; therefore such a manifold is said to be globally orientable.

The most striking differences between one manifold and another are global differences, differences that can be discerned only by studying entire atlases, not single charts. One global property of a surface is orientability. For example, the atlases for a torus and for the surface known as a Klein bottle are quite similar, but the Klein bottle has two peculiar features that set it quite apart from the torus. First, the Klein bottle cannot be constructed in space without intersecting itself in places where it should not intersect itself. Second, it cannot be oriented in space. Self-intersection cannot be inferred directly from the atlas. It is an extrinsic feature. Nonorientability is an intrinsic feature, however, and it can be discovered by following a moving clock face from one chart in the atlas to the next. Nonorientability is a global property. A second global property is connectivity, arising from the question of whether every closed loop sliced on a surface separates the surface into two pieces. A sphere and a torus differ in their connectivity. There are analogous notions of connectivity for higher-dimensional manifolds. The study of global properties is a part of topology. It thus becomes evident that in a manifold—including Einstein's three-sphere model of space—a number of small localities having a quite ordinary and familiar geometry can be combined to produce a global effect that is both novel and surprising. This aspect of manifolds is also a part of the fascination of the drawings of Maurits C. Escher.

In the search for the origins of Einstein's ideas much attention has been given to the non-Euclidean geometry that was developed in the early 1800's. That emphasis is somewhat misleading. Several mathematicians, including Johann Heinrich Lambert, who lived between 1728 and 1777 and who was a friend of Kant's, had come to the conclusion that a different collection of theorems, logically as sound as those of Euclidean geometry, could be derived from a new system of axioms that differed slightly from Euclid's. A smaller number of 19th-century mathematicians, notably Gauss, János Bolyai and Nikolai Lobachevski, saw further that such a new system could plausibly describe the structure of physical space just as well as Euclid's. It was no longer possible to maintain Kant's position that Euclidean geometry was synthetic a priori. The antinomy of space persisted, however, because the new Lobachevskian space (as it is often called) had the same overall topological structure as Euclidean space. In particular any finite collection of galaxies in it would still have to have a boundary.

The 19th century was a time of tremendous developments in geometry. By the 1870's comprehensive systems embracing all Euclidean, non-Euclidean and projective geometry had evolved. Curiously, the question of whether the geometric systems were physically relevant drifted from the center of attention and even became rather confused. Henri Poincaré, one of the most eminent mathematicians of the time, even maintained that there was no strictly correct geometry for the description of space. The choice of one geometry instead of another was purely a matter of convention, he said, or was no more consequential than the choice between Fahrenheit and centigrade thermometers to measure temperature.

The mathematics of Einstein's general theory of relativity is quite distinct in style from that of the prevailing systematic geometries at the beginning of the 20th century. It is a part of differential geometry, which, as the name suggests, exploits the power of calculus. With it Einstein showed how to interpret gravitation as being a curvature of space. In other words, distance distortions will be present in a model of even a small portion of space if the space contains an appreciable amount of matter. That distortion, and not gravitational "force," then dictates the paths of moving bodies. Curvature is therefore a local phenomenon as well as a global one. In fact, the general theory of relativity is mainly occupied with the local consequences of curvature. It is difficult to imagine how the geometry of the general theory of relativity, inextricably bound up with matter as it is, could have evolved from the tidy axiomatic systems of Euclid and Lobachevski. Borrowing a locution from the world of the New York theater, where plays can be presented off-off-Broadway, we could say that the geometry Einstein used is non-non-Euclidean. Einstein's ideas were cast in a language very different from even non-Euclidean geometry, called the absolute differential calculus. Until Einstein used it and changed its name to tensor analysis, it had the reputation of being the kind of pure mathematics that had no connection with the real world.

Ironically tensor analysis did not start that way. Its origins can be found in the work of Bernhard Riemann (1826–1866). On the occasion of Riemann's appointment to the faculty of the University of Göttingen in 1854 at the age of 27 he opened an entirely new line of thinking in his probationary lecture, titled *On the Hypotheses That Lie at the Foundations of Geometry*. The topic had been selected for him by his teacher and colleague, Gauss. The entire modern viewpoint can be found in Riemann's lecture: the concept of an n-dimensional manifold; the study of a manifold's intrinsic geometry—particularly curvature—by extending Gauss's work on surfaces; even the radical notion that geometry and physics are inseparable, that is, that the presence of matter determines the curvature of space. Riemann's central concern, however, was with the implications his ideas had for the structure of physical space. His own words (translated) explain it well:

"That space is an unbounded three-dimensional manifold is an assumption that is employed for every apprehension of the external world, by which at every moment the domain of actual perceptions is filled in and the possible locations of a sought-for object are constructed; in these applications it is continually confirmed. For that reason the unboundedness of space has a greater

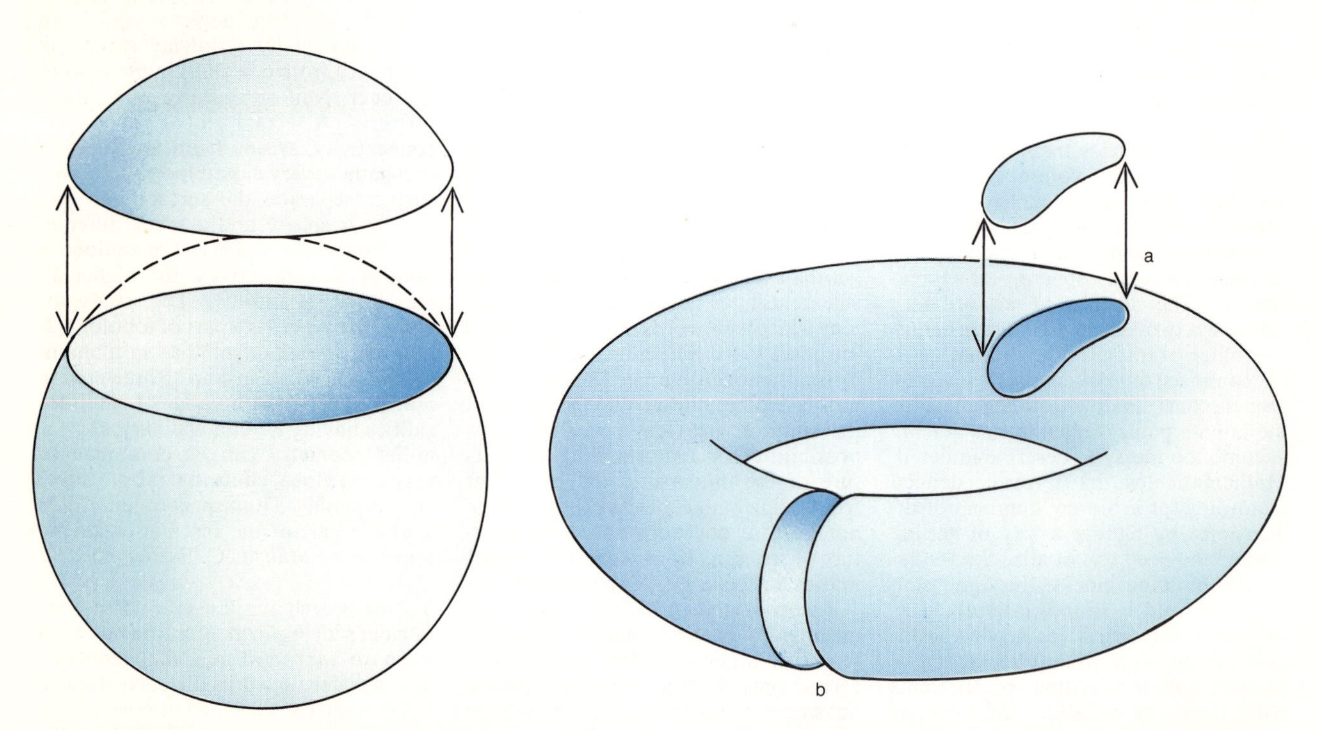

CONNECTIVITY is another global property of a surface, and it refers to the question of whether or not every closed loop on a surface separates the surface into two pieces. For example, the connectivity of a sphere and that of a torus are different. Any loop on a sphere separates it into two pieces (*left*). Such loops (*a*) also exist on the torus (*right*), but there are certain loops (*b*) that do not separate it. The fact that nonseparating loops can exist for some surfaces and not for others reflects a basic topological difference between those surfaces.

certainty than any external experience. But its infinitude in no way follows from this. On the contrary, space would necessarily be finite if we assumed that bodies exist independently of position—so that we could ascribe constant curvature to space—as long as this curvature had a positive value, however small."

Riemann treated space as a manifold, and he recognized that one of its possible structures is given by what we have been calling Einstein's three-sphere. Thus it was Riemann who first saw that a finite unbounded universe was conceptually possible. He, not Einstein, actually provided the way around Kant's antinomy of space.

Although Gauss, for one, was astonished by what he heard Riemann say, the wider scientific community did not become generally aware of Riemann's lecture until it was published posthumously in 1868. (Riemann died of tuberculosis at 39.) During the next half-century Riemann's mathematical ideas were developed extensively, but their applications to space were virtually forgotten. A generation later Einstein apparently rediscovered the full wealth of Riemann's ideas about the physical world. Einstein's work, however, is no mere copy of what Riemann had already done. The general theory of relativity is above all a physical theory, a coherent and detailed account of the underlying geometric character of gravitation. Riemann provided the geometric language, but Einstein's physics was radically new. Riemann had only hinted at it.

The positions of Einstein and Kant are by no means antithetical. On the contrary, as we have seen, one of the fundamental tenets of Kant's metaphysical idealism is that space is not a thing but one of the forms through which we organize our perceptions of things. Moreover, the structure of our perceptual organization is given a priori. Those thoughts of Kant's are implicitly accepted by relativity. The basic quarrel between Einstein and Kant is over the structure of space. Kant assumed, partly because he saw no more general possibility, that space must be Euclidean; Einstein maintained that it is Riemannian, and he includes Kant's position as a special case. That explains how the antinomy of space could arise in the first place, and also how Riemann and Einstein could resolve it without completely undermining Kant.

"WATERFALL," a black-and-white lithograph by Maurits C. Escher, is an example from the world of art of how space might be locally quite ordinary and yet globally quite surprising.

A final word about Dante, following some observations made by Andreas Speiser in his book *Klassische Stücke der Mathematik*. In the first two books of *The Divine Comedy* Dante traverses the material world from the icy core of the earth, the abode of Lucifer, to the Mount of Purgatory. In the last book, *Paradise*, Dante's beloved Beatrice guides him up through the nine heavenly spheres, each sphere larger and more rapidly turning than the last, until he reaches the Primum Mobile, the ninth and largest sphere and the boundary of space. His goal is to see the Empyrean, the abode of God. It finally appears to him, as a blinding point of light surrounded by nine concentric spheres that represent the angelic orders responsible for the motions of the material spheres. Dante is puzzled, however, because the smaller the radius of each Empyrean sphere is, the faster the sphere turns. Beatrice explains that there is no paradox between the material and the spiritual spheres; every sphere, whether material or spiritual, turns faster the more perfect or divine it is. The spiritual world completes the material world exactly as one viewing screen of Einstein's model of the galactic system completes the other. The overlap between the two is revealed by the correspondence between the celestial spheres and their angelic counterparts, and again as in Einstein's model the farther a sphere is from the center of one chart, the nearer its counterpart is to the center of the other. And the speeds of the material spheres and of the spiritual spheres are in harmony.

Speiser suggests that Dante was able to come to this remarkable vision because his geometric knowledge was derived from astronomy and not from Euclid, which he scarcely knew. Here is a translation by Barbara Reynolds (Penguin, 1962) of the conversation between Dante and Beatrice in Canto XXVIII:

About this Point a fiery circle
whirled,

With such rapidity it had outraced
The swiftest sphere revolving round the world.

This by another circle was embraced,
This by a third, which yet a fourth enclosed;
Round this a fifth, round that a sixth I traced.

Beyond, the seventh was so wide disclosed
That Iris, to enfold it, were too small,
Her rainbow a full circle being supposed.

So too the eighth and ninth; and each and all
More slowly turned as they were more removed
Numerically from the integral.

Purest in flame the inmost circle proved,
Being nearest the Pure Spark, or so I venture,
Most clearly with Its truth it is engrooved.

Observing wonder in my every feature,
My Lady told me what I set below:
"From this Point hang the heavens and all nature.

Behold the circle nearest it and know
It owes its rapid movement to the spur
Of burning love which keeps it whirling so."

"If manifested in these circles were
The cosmic order of the universe,
I should be well content," I answered her;

"But in the world below it's the reverse,
Each sphere with God's own love being more instilled
The further from its centre it appears;

Whence, if my longing is to be fulfilled,
Here in this wondrous and angelic fane,
Where love and light alone the confines build,

I must entreat thee further to explain
Why copy from its pattern goes awry,
For on my own I ponder it in vain."

"There's naught to marvel at, if to untie
This tangled knot thy fingers are unfit,
So tight 'tis grown for lack of will to try."

Then she went on: "This is no meagre bit
I'll give to thee. Wouldst thou be filled? Then take,
And round its content ply thy subtle wit.

Material circles in the heavens make
Their courses, wide or small, as more or less,
Through all their parts, of virtue they partake.

The greater good makes greater blessedness;
More blessedness more matter must enclose,
If all its parts have equal perfectness.

It follows that the sphere, which as it goes
Turns all the world along, must correspond
To this, the inmost, which most loves and knows.

Hence, if thou wilt but cast thy measure round
The angels' *power*, not their circumference
As it appears to thee, it will be found

That wondrous is the perfect congruence
Which every heaven with every mover shows
Between their corresponding measurements."

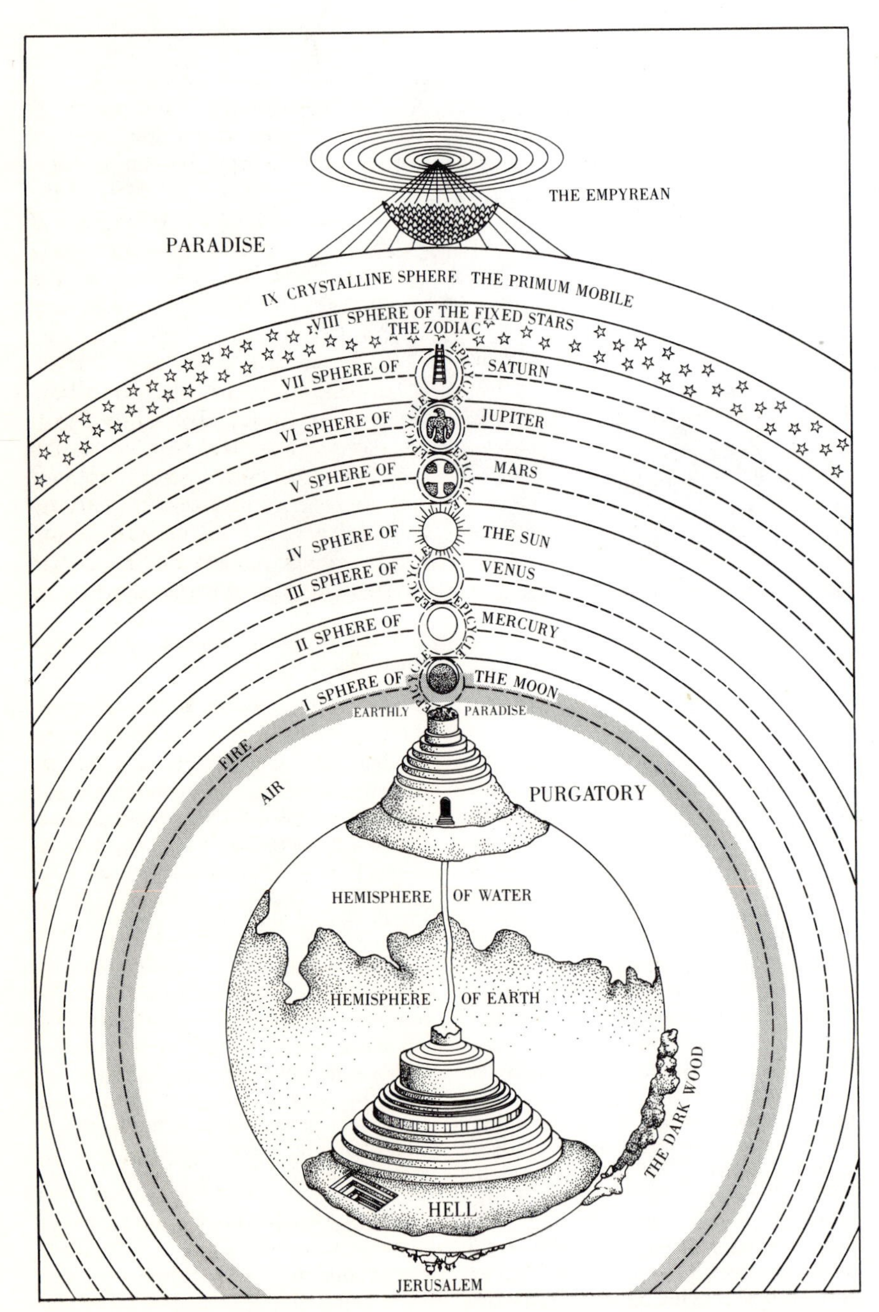

DANTE ALIGHIERI'S SCHEME OF THE UNIVERSE in illustration from "Paradise" in *The Divine Comedy* extends Aristotelian cosmology in a modern way. It is discussed in text.

Cosmology before and after Quasars

4

by Dennis Sciama
September 1967

A review of *The Measure of the Universe,* by J. D. North. Oxford University Press, 1965.

I have often wondered what it must have been like to be a nuclear physicist in the early 1930's, particularly in 1932—that *annus mirabilis* which saw the discovery of the neutron and the positron and the first splitting of the nucleus by artificially accelerated particles. Now I think I know. As a cosmologist I have seen in the 1960's a similar stream of discoveries following one on another at an almost indecent rate. The years 1963 to 1965 stand out, beginning with the discovery of quasars, followed by the measurement of the fantastic red shifts possessed by some of them and culminating in what is perhaps the greatest discovery of them all: the cosmic black-body radiation. I should say at once that the evidence is not yet decisive that any of these discoveries has cosmological significance, but it is good enough to have reduced most cosmologists, who are traditionally starved for basic observations, to a state of bewildered euphoria.

These reflections are prompted by the publication of an interesting book called *The Measure of the Universe,* written by J. D. North, an Oxford philosopher. It is a history of modern cosmology that ends just before the new period begins. It barely mentions quasars and does not mention the cosmic black-body radiation at all. This is no criticism, because the book was written too early for it to have done so. It was therefore written at the right time to take stock of the first great period in cosmology. That period, which had only the expansion of the universe to explain, we might justly call the geometrical period. Today we are well and truly launched into the astrophysical period.

To be fair to the early theorists, they did predict the expansion of the universe before it was discovered. By the early 1920's it was clear to Willem de Sitter, Alexander Friedmann and Hermann Weyl that Einstein's field equations of general relativity had as solutions homogeneous and isotropic model universes whose material substratum was in a state of expansion, the relative velocity at which two particles moved apart simply being proportional to their distance (except for refinements for very widely separated particles). Moreover, if the debatable cosmical constant was dropped from the field equations, as Einstein later urged, then *all* the homogeneous and isotropic solutions exhibited expansion (or contraction if one cared to reverse the sense of time). It was not until 1929 that the Hubble law, that the observed red shift is proportional to the distance of a galaxy (as estimated by various more or less dubious criteria), was first stated.

North's book gives a thorough account of this classical phase of theoretical cosmology. There were many controversies at the time about the properties of the various models. That phase is now over, and the correct results are enshrined in standard theory. There were also controversies of a different nature; not everyone accepted the view that general relativity was uniquely fitted to deal with the universe as a whole. Various "heretical" theories were proposed, notably by Sir Arthur Eddington, E. A. Milne, P. A. M. Dirac and Pascual Jordan, and they are described too. I deliberately mention separately the steady-state theory of Herman Bondi, Thomas Gold and Fred Hoyle, because I think it is fair to say that of all the heretical theories this is the one that has irritated and excited the most people, has provoked the most good astrophysics and has more or less survived to the present day.

I say "more or less" because one of the consequences of the new turn of events—of cosmology becoming astrophysical—is that if the red shifts of the quasars are cosmological in origin, and if the universe is filled with black-body radiation, then the chances of the steady-state theory surviving are very small indeed. I want to make clear why this is so, and to discuss what further information we can hope to extract from the new results and their likely future extensions. I must add that for me the loss of the steady-state theory has been a cause of great sadness. The steady-state theory has a sweep and beauty that for some unaccountable reason the architect of the universe appears to have overlooked. The universe in fact is a botched job, but I suppose we shall have to make the best of it.

One of the botches is the existence of a singularity, that is, a moment when the density of the universe was infinite. To be more precise, this is what general relativity requires for the homogeneous and isotropic models to which I have referred. It has sometimes been suggested that the singularity would go away as soon as one admitted that the real universe was neither exactly homogeneous nor exactly isotropic; in such circumstances the galaxies would not move quite radially, and so the matter they are made of would not all have emerged from exactly one point in the past. It has recently been shown by Stephen Hawking and others, however, that the orthodox theory of general relativity, without on the one hand a cosmical constant and on the other assumptions of exact symmetry, still requires the physical properties of the universe to have been singular at some time or times.

It was to avoid such an unpleasant singularity (and for other reasons too) that Bondi, Gold and Hoyle proposed in 1948 a deviation from orthodox general relativity that would allow the continual creation of matter at a rate just compensating for the expansion of the universe. The resulting mean density of the universe (and indeed all its other average properties) would then be independent of time, leading to a steady state that would automatically persist forever. It is this magnificent conception we must now reluctantly abandon.

The first evidence against the steady-state theory came from counts of celestial radio sources, conducted notably by Sir Martin Ryle and his colleagues in Britain, but also by B. Y. Mills and J. G. Bolton in Australia and by M. Ceccarelli in Italy. These counts showed that the number of faint radio sources was far too large compared with the number of bright sources to be compatible with the steady-state theory. This evidence has given rise to much controversy, mainly because the majority of sources concerned have not yet been identified optically. Accordingly inferences drawn from these counts have been surrounded by an aura of uncertainty. A straightforward interpretation, stressed by Ryle and studied in detail by William Davidson and by Malcolm Longair, requires that the radio sources exhibit intrinsic evolution. That is to say, the faint sources, which are mostly at great distances and so are now being seen as they were a long time in the past (because of the time their radio waves take to reach us), must have average properties different from the bright sources, which are mostly relatively near and so are being seen almost contemporaneously with ourselves. Such evolution is of course incompatible with a steady-state universe, but it would be expected in a universe evolving from a dense state to a dilute one, a universe to which one can attach the concept of an age.

Needless to say, there have been several implausible attempts to evade Ryle's argument. My own attempt has turned out to be correct, but not in the way I intended. I proposed before the discovery of quasars that the radio sources in Ryle's catalogue consisted of two different populations. One population was to be the radio galaxies that had already been identified optically and were well known. For the second population I proposed the existence of radio stars in our galaxy, whose distribution between bright and faint sources would explain the anomalous counts but would have nothing to do with cosmology. It has turned out that a second population of starlike (that is, unresolvable) radio sources does exist. Moreover, they are just the ones responsible for the excess of faint sources (as has been shown independently by Philippe Véron and Longair). But these quasi-stellar radio sources, or quasars, have large red shifts and are therefore not the objects I had in mind.

It is true that a few physicists and astronomers (James Terrell, Geoffrey and Margaret Burbidge, Hoyle) hold, with differing degrees of assurance, that these large red shifts are not cosmological in origin, and that the quasars are within, or relatively close to, our galaxy. This would be in tune with my proposal, but I find their arguments unconvincing. If the red shifts have a Doppler origin, that is, if the quasars are receding from us rapidly as a result of a local explosion, the question arises of why we do not see any blue shifts from quasars fleeing from neighboring galaxies toward us. Of course, if quasar emission is a sufficiently rare process, the nearest such galaxies would be too far away for their quasars to be visible, but then why should we be privileged to witness such a rare event so close to us? Clearly this is possible but unlikely.

On the other hand, if the red shifts do not have a Doppler origin but arise, say, from the Einstein gravitational effect, and if the sources are distributed uniformly in space with the ones observed so far quite close to us, then we would not expect the source counts to manifest an excess of faint objects. The relative number of bright and faint sources should be the same as if there were no red shift (that is, the same as for a uniform distribution of stationary sources) and this is not what is observed. We conclude that the red shifts are most probably cosmological in origin. On this basis Martin Rees and I have carried out an analysis of the red shifts of the quasars, and we find again that there are too many faint sources with large red shifts to be compatible with the steady-state theory.

In weighing the significance of what I have said so far it is important to understand how accidental it is that we should be able to observe such large red shifts so easily. Objects with these large red shifts are so distant that the different cosmological theories make substantially different predictions about them, but such objects are visible only because quasars happen to be a hundred times brighter than galaxies. In contrast, the existence of cosmic black-body radiation, which also serves to distinguish among different theories, is intimately bound up with the development of the universe itself.

The detection in 1965 of excess radiation at microwave frequencies (that is, at wavelengths of a few centimeters) and the evidence that it has a black-body spectrum (that is, is in thermal equilibrium characterized by a single temperature) has been described in "The Primeval Fireball," by P. J. E. Peebles and David T. Wilkinson [*see Scientific American* June 1967]. The temperature observed is about 3 degrees absolute. I should like to make the following comments on this result:

1. No plausible noncosmological explanation has yet been proposed (and not for want of trying).

2. A natural cosmological explanation does exist if the universe was once very dense.

3. A temperature significantly greater than 3 degrees would not be compatible with our general knowledge of radio astronomy and high-energy astrophysics.

I shall say no more about the first point, but I should like to discuss the second and third in a little more detail. As in the case of the expansion of the universe, the existence of cosmic black-body radiation was predicted before it was observed. Around 1950 it was proposed by George Gamow and his associates that the early, dense stages of the universe were very hot, a state of affairs often described as the "hot big bang." Their reason for making this proposal was that in such conditions thermonuclear reactions could occur at an appreciable rate, converting primordial hydrogen into helium and possibly heavier elements. By choosing the right early conditions Gamow was able to account approximately for the abundance of helium with respect to hydrogen that is observed today. This helium problem is actually in a very confused state at the moment, but the important point here is that if the early, dense stages were hot, unquestionably there was ample time for matter and radiation to come into thermal equilibrium. At that time, then, the radiation would have had a black-body spectrum. Moreover, at all times thereafter the spectrum would remain that of a black body, the radiation simply cooling down as the universe expanded. Gamow's original calculations of helium formation led him to predict for the present temperature of the black-body radiation a value of about

30 degrees absolute, but modern calculations are compatible with a lower temperature, in particular with a temperature of 3 degrees absolute.

As I have mentioned in my third point, we now know that a temperature as high as 30 degrees can be ruled out. Cosmic ray protons and electrons interacting with such radiation would produce effects that could be observed, and they are not. Three degrees is about the highest permitted temperature from this point of view, and 3 degrees is just what has been found. The connection between the microwave observations and Gamow's theory was made by Robert H. Dicke and his colleagues at Princeton University. In fact, they had the bad luck to be setting up apparatus to look for the excess radiation when it was discovered accidentally down the road by Arno A. Penzias and Robert W. Wilson of the Bell Telephone Laboratories.

Can the steady-state theory account for the excess radiation? It would be reasonable to propose that along with the newly created matter there comes into existence newly created radiation; indeed, some such effect would be expected as a result of the creation process itself. But why the observed spectrum should be that of a black body over a wide range of wavelengths is totally obscure. It is therefore critically important to establish without doubt that the actual spectrum is that of a black body. The present evidence is strong but not decisive. Further work is being done, and this point should be settled fairly soon.

There is one final property of the radiation that I want to discuss because it is in some ways its most exciting feature, and that is its degree of isotropy—its uniformity with respect to direction of arrival. Just a few months ago R. B. Partridge and Wilkinson announced that any anisotropy is less than a few tenths of a percent. I have heard cynical scientists comment that this result throws doubt on the whole phenomenon, on the grounds that noise generated internally by the observing instruments would be more "isotropic" than externally generated noise, even noise coming from a highly isotropic universe. To silence this cynicism it is necessary to show that the universe is likely to be isotropic to the required degree. A first step in this direction has recently been taken by Charles W. Misner, who has shown that for a certain class of model universes any initial anisotropy would be rapidly removed by a rather exotic form of viscosity involving the pairs of neutrinos that would be excited by the high temperatures then prevailing. Misner's program is to allow the universe to start out as irregularly as it wishes and then to show that all irregularities would be damped out by the action of accepted physical processes, except for those irregularities we actually observe (such as clusters of galaxies).

Another intriguing aspect of the isotropy measurements is that they can be used to determine our "absolute" velocity, that is, our velocity with respect to the distant matter that last effectively scattered the radiation. Because of the Doppler effect such a velocity would reveal itself by leading to a slightly higher temperature for the radiation ahead of us and a slightly lower temperature for the radiation behind us. The present limit on the anisotropy corresponds to a velocity limit of 300 or 400 kilometers per second. One contribution to the expected velocity comes from the sun's known rotation around our galaxy of about 250 kilometers per second. Even this, however, depends on the correctness of Mach's principle [see "Inertia," by Dennis Sciama; SCIENTIFIC AMERICAN, February, 1957]. According to Mach's principle, the local nonrotating frame of reference as determined dynamically coincides with the frame in which distant matter is not rotating. The well-known rotation of our galaxy, with which is associated the galaxy's dynamical flattening, would then be a rotation with respect to distant matter, and therefore to the effective sources of the black-body radiation.

To obtain our net motion relative to these sources, however, we must also allow for the peculiar motion of our galaxy in the local group of galaxies (which has been estimated to be about 100 kilometers per second) and a possible systematic motion of our galaxy in the local supercluster. This supercluster is believed by some astronomers (Vera Rubin, Gérard de Vaucouleurs, K. F. Ogorodnikov) to be a flattened system of galaxies rotating around the Virgo cluster. I have recently rediscussed our possible motion in the supercluster and arrived at a tentative rough estimate for our net motion through the black-body radiation of about 400 kilometers per second in the general direction of the center of our galaxy.

Future observations of the black-body radiation should be able to test this prediction, and in view of the uncertainty surrounding the notion of a local supercluster I would not be at all surprised to find that it is wrong. My point is simply that yet another new range of problems has been opened up for the cosmologist by the existence of the black-body radiation. We have come a long way in a few years from the geometrical considerations described in North's book, and we can rejoice. Cosmology has at last become a science.

5 The Cosmic Background Radiation

by Adrian Webster
August 1974

The space between the galaxies is filled with radiation ranging from radio waves to gamma rays. The radiation has been generated by various processes, some of which are traced to the "big bang"

No part of the universe is empty. The space between the planets contains the "wind" of ionized gas expelled by the sun and the dust that is seen from the earth as the zodiacal light. The space between the stars contains a variety of materials, ranging from the hydrogen whose emission and absorption at the wavelength of 21 centimeters is studied by radio astronomers to the dust that weakens and reddens the light of distant stars. Even on the largest scale of all, the vast reaches of space between the galaxies are not empty. To be sure, no gas or dust or any other form of matter has been detected there, but it is quite clear that the whole of that space is permeated by a uniform background of electromagnetic radiation. This cosmic background radiation has now been detected across most of the electromagnetic spectrum, from radio waves at a wavelength of 300 meters to gamma rays at a wavelength of 10^{-14} meter. It provides a wealth of information on the history of the universe back to its origin in the "big bang."

The cosmic background radiation has been measured only within the past decade, but interest in the subject goes back two and a half centuries. Early in the 18th century Edmund Halley asked: "Why is the sky dark at night?" This apparently naïve question is not easy to answer, because if the universe had the simplest imaginable structure on the largest possible scale, the background radiation of the sky would be intense. Imagine a static infinite universe, that is, a universe of infinite size in which the stars and galaxies are stationary with respect to one another. A line of sight in any direction will ultimately cross the surface of a star, and the sky should appear to be made up of overlapping stellar disks. The apparent brightness of a star's surface is independent of its distance, so that everywhere the sky should be as bright as the surface of an average star. Since the sun is an average star, the entire sky, day and night, should be about as bright as the surface of the sun. The fact that it is not was later characterized as Olbers' paradox (after the 18th-century German astronomer Heinrich Olbers). The paradox applies not only to starlight but also to all other regions of the electromagnetic spectrum. It indicates that there is something fundamentally wrong with the model of a static infinite universe, but it does not specify what.

Olbers' paradox was resolved in 1929, when Edwin P. Hubble showed that the universe is not static but is uniformly expanding. The galaxies are all receding from one another, and the velocity of recession, as it is perceived on the earth, is directly proportional to the galaxy's distance. The velocity of recession has a strong effect on the light traveling from the galaxy to the earth. First of all, with each passing moment the successive photons (quanta of light) emitted by the stars in the galaxy have farther to go in order to reach the earth, so that their rate of arrival is lower than it would have been if the galaxy had been stationary. Second, the Doppler effect shifts the photons to lower frequencies, so that they have less energy.

Together these two effects weaken the light from the stars in a distant galaxy over and above the dimming due to the galaxy's distance. Both effects become particularly strong when the speed of the galaxy is a substantial fraction of the speed of light, because at that point the special theory of relativity must be taken into account. The result of all these weakening effects is that the energy density of starlight does not reach enormous values and cause the sky to be as bright as the sun. The same argument applies to photons of all other wavelengths.

At this writing the background radiation has been observed in four regions of the electromagnetic spectrum: the radio region, the microwave region, the X-ray region and the gamma-ray region. The background radiation at each of the different wavelengths provides a different kind of information on the history of the universe, so that I shall deal with each region separately, starting at the long wavelengths and continuing toward the short.

The cosmic background radiation in the radio region is detected between the frequencies of one megahertz (million cycles per second) and 500 MHz, corresponding to the wavelengths between 300 meters and 60 centimeters. This radiation is somewhat difficult to detect and measure because our galaxy is itself a source of radio waves. The background flux of radio power from all directions is dominated by this foreground of galactic radiation. Measurements of the brightness of the radio sky at the frequency of 20 MHz in the direction of the Large Cloud of Magellan have shown, however, that some of the radio power comes from outside our galaxy. Within the Large Magellanic Cloud, which is one of two small galaxies close to ours, is a cloud of ionized gas that absorbs radio waves at the frequency of 20 MHz. C. A. Shain of the Commonwealth Scientific and Industrial Research Organization in Australia has found that in the direction of this cloud of ionized gas, designated 30 Doradus, there is a decrease in the radio brightness of the sky. His observations show clearly that the ionized gas of 30 Doradus is absorbing radiation, so that there must be radiation coming from a distance greater than that of the Large Magellanic Cloud and therefore coming from well outside our galaxy.

To study the background radiation over a wide range of frequencies in the radio region a technique developed by radio astronomers at the University of Cambridge is used. Accurate maps of the radio sky are made at a variety of wavelengths and are compared with one another. The background radiation does not vary appreciably in strength between one direction and another, and so it has the effect of increasing all the measurements at one wavelength on any map by a constant amount; the zero level to which the brightness is referred on each map is established by the background radiation. The way in which the zero level varies from map to map at various wavelengths gives the strength of the background radiation at each wavelength.

This mapping technique has revealed that the intensity, or energy density, of the radio background drops off as the spectral frequency increases. That is, the amount of energy at shorter wavelengths is less than that at longer ones. This type of spectrum is commonly encountered in radio astronomy; it is called a nonthermal spectrum to distinguish it from the thermal radiation of a hot gas, where the relation between intensity and frequency is totally different. The nonthermal radiation is believed to be generated by the synchrotron process, in which high-energy charged particles spiraling along the lines of force in a magnetic field emit energy at radio wavelengths.

The origin of the radio background is

CHART OF THE RADIO SOURCES in a small patch of the sky was made by the three dishes of the One Mile Telescope at the University of Cambridge, which are arranged as an interferometer. Each set of peaks on the chart is a radio galaxy or a quasar. Charts such as this one show that there are more faint celestial radio sources than one would expect from the number of bright ones. It is believed that the cosmic background radiation in the radio region of the spectrum is due to all the radio sources in the sky added together. The faint sources are actually powerful sources at great distances, seen now as they were in the distant past because of the time taken for the signals to reach the earth. Such observations indicate sources were more numerous in the past than they are now.

well understood: it is the emission of all the radio sources in the universe added together. These sources, such as radio galaxies and quasars, have nonthermal spectra of just the same kind as the background radiation. It is possible to count enough radio sources on maps made with radio telescopes of the highest sensitivity to infer that the radiation from all of them, including sources too faint to detect individually, adds up to the strength of the radio background. By analyzing the radio background one can study all at once many more radio sources than one could ever hope to detect and examine individually. A great deal of useful information has emerged from such analysis.

For example, the radio background has confirmed evidence from surveys of individual radio sources that there were many more radio galaxies and quasars in the recent cosmic past than there are now. The reason is that if one were to calculate the strength of the background from the present number of radio galaxies and quasars, the result would be much smaller than the observed strength. Allowing for a greater number of these objects in the past makes the calculations agree with the observations. Such confirmation is of vital importance to modern cosmology because it shows, among other things, that the steady-state theory of the universe is quite wrong. The steady-state theory posits that any one part of the universe is on the average much the same as any other part, and that at present the universe is much the same as it always was and ever will be. One consequence of the steady-state theory is that it calls for the continuous creation of matter to counteract the decrease in density caused by the universe's expansion. The steady-state theory neatly sidesteps all problems about the origin of the universe and the beginning of time by stating that there need not have been a beginning. It is clear, however, that since there were many more radio sources in the past than there are now, the universe was not the same then, and that the fundamental postulate of the steady-state theory is false.

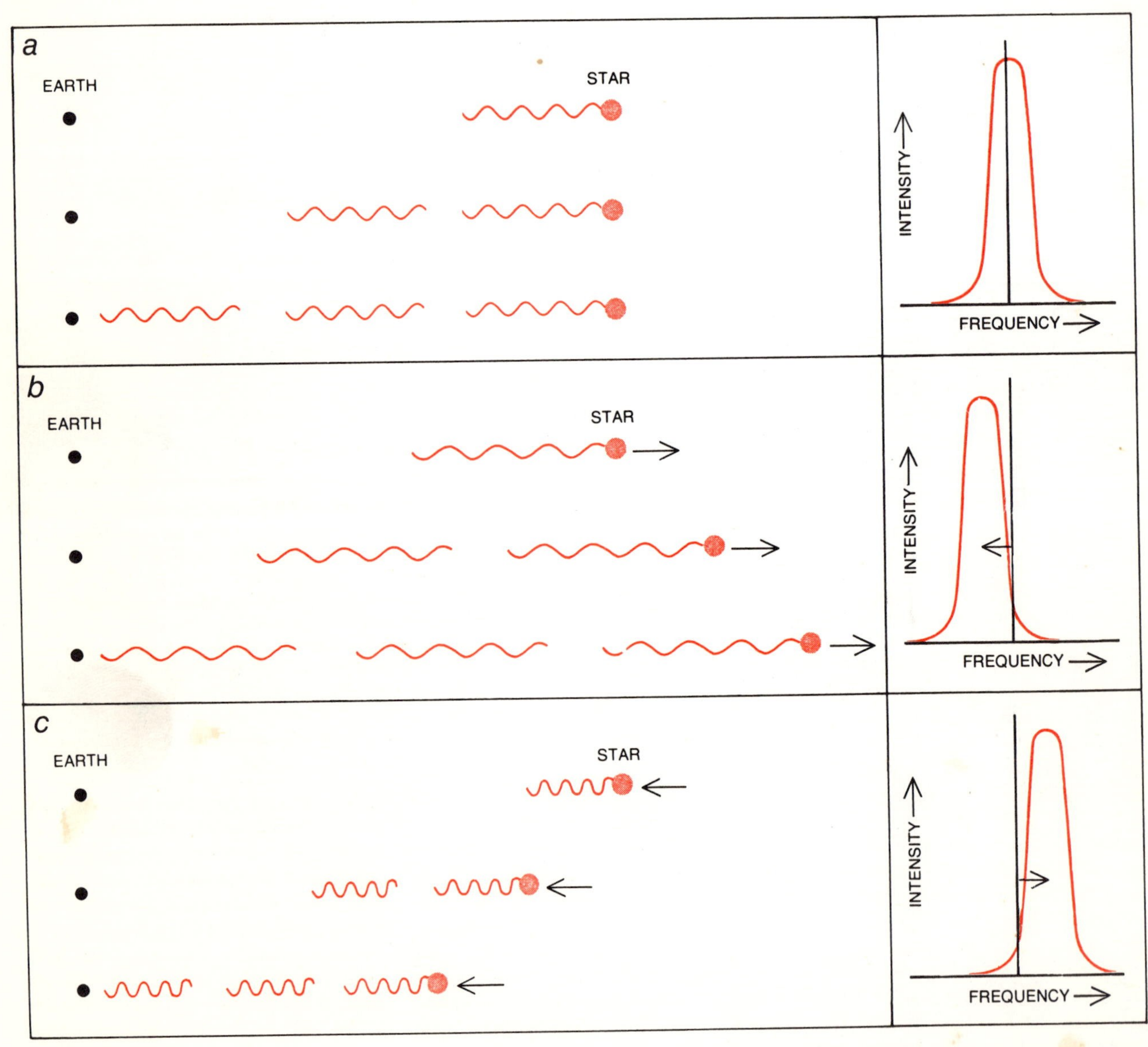

DOPPLER SHIFT of the light from a moving star can be used to determine both the velocity and the direction of the star with respect to the earth. If the star is stationary (*a*), its light will be unaffected, and each of the lines in its spectrum will be at the "rest" frequency (*indicated by the vertical line in the spectrum at right*). If the star is moving away from the earth (*b*), its light will be shifted toward longer (redder) wavelengths and will also be dimmed because fewer photons (*wavy line segments*) emitted by the star will reach the earth within a given time interval than if the star were at rest. Conversely, if the star is moving toward the earth (*c*), its light will be shifted toward shorter (bluer) wavelengths and the photons will be arriving at the earth with greater frequency.

The alternative hypothesis to the steady-state theory is the big-bang theory. This theory states that at the beginning of time all the matter in the universe exploded out of a superdense kernel. The observations of the radio background are easy enough to fit into the big-bang theory: the era from about a quarter of the age of the universe up to the present was the era of radio galaxies and quasars. That era is apparently coming to an end, so that there are few of these objects left. In the future there may well be none at all.

One feature of the radio background is conspicuous by its absence. It is well known that in the space between the stars of our galaxy there is a substantial amount of neutral (un-ionized) hydrogen radiating at its characteristic wavelength of 21 centimeters. The 21-centimeter emission has been sought in the spectrum of the cosmic background radiation and has not been found. Consequently there cannot be much neutral hydrogen in the space between the galaxies. This result is of particular interest because according to the canonical hot-big-bang theory, to which I shall return below, the universe must have been full of neutral hydrogen when the initial fireball of the big bang had cooled off. Presumably much of the hydrogen has been used up in the building of stars and galaxies, and the rest may have been reionized.

The next region of the spectrum in which the cosmic background radiation has been detected is the microwave region. The microwave background predominates at all frequencies between 500 MHz and 500 GHz (gigahertz, or 10^9 cycles per second), corresponding to wavelengths between 60 centimeters and .6 millimeter. The microwave background was discovered in 1965 by Arno A. Penzias and Robert W. Wilson of Bell Laboratories [see "The Primeval Fireball," by P. J. E. Peebles and David T. Wilkinson; SCIENTIFIC AMERICAN, June, 1967]. The early measurements of the intensity of the radiation at various wavelengths produced a curve resembling that of a "black body" (an ideal radiator) at a temperature close to 3 degrees Kelvin (degrees Celsius above absolute zero). The radiation seemed to be constant in strength across the sky, and its existence apparently fitted perfectly into the framework of the big-bang theory.

Since that time many more measurements of the microwave background have been made. All have confirmed its black-body curve between 408 MHz and 115 GHz, corresponding to wavelengths between 74 centimeters and 2.6 millimeters. The best and most recent value for the characteristic temperature of the radiation is 2.76 degrees K.; the measurements are so accurate that it is unlikely that this figure differs from the true one by more than about 3 percent. Instruments carried aloft by balloons and rockets should soon yield accurate measurements in the highest range of microwave frequencies, between 115 GHz and 500 GHz, corresponding to wavelengths between 2.6 millimeters and .6 millimeter. These measurements must be made from above most of the earth's atmosphere because the atmosphere strongly absorbs radiation of millimeter wavelengths. They are very important measurements because they will provide the final check on whether or not the microwave curve is truly a black-body one.

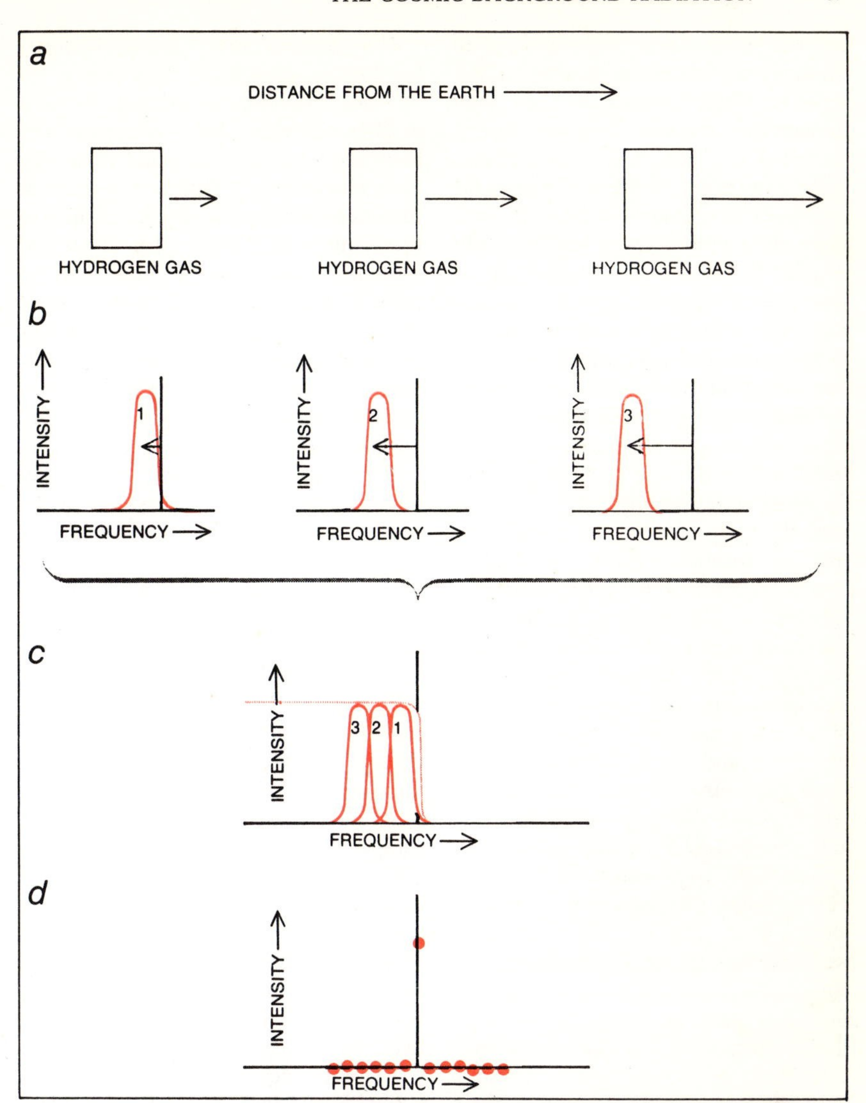

NEUTRAL (UN-IONIZED) HYDROGEN GAS does not fill the space between the galaxies, at least insofar as it has not been detected by radio telescopes. If the universe were filled with hydrogen, presumably the gas that is farther from the earth would be receding faster than the gas that is closer (*a*), just as the galaxies are observed to recede. Therefore the emission line at the radio wavelength of 21 centimeters (corresponding to a frequency of 1,420 megahertz) of the gas farther from the earth (*b*) would be Doppler-shifted more toward longer wavelengths and lower frequencies than the emission line of the gas closer to the earth would be. The result would be that all the shifted emission lines from the gas at the various distances would add up (*c*) to produce a "step" (*light color*) in the radio spectrum below 1,420 MHz. No such step, however, has ever been found (*dots in color*) in the radio spectrum (*d*). The clear signal at exactly 1,420 MHz is from gas in our own galaxy.

The canonical hot-big-bang theory is an outline that has recently been developed to account for the overall history of the universe. It has been deduced by starting at the present condition of the universe and working backward in time, using all the known laws of physics to calculate what the main processes were at each stage. The details of the calculations are fascinating; it is quite surprising how much can be reliably deduced about what went on so long ago. Here, however, there is room only for a brief summary.

The illustration on page 41 charts the history of the universe from the time when its overall temperature was 10^{12} degrees K. Before that time the universe was full of short-lived exotic particles and antiparticles at tremendous density, temperature and pressure. These particles were in equilibrium with the field of radiation, that is, the particles could interact to produce photons and the photons could interact to produce particles. The higher the energy of the photons was, the more massive and peculiar the particles were that each photon could create. By the time the temperature had dropped to 10^{12} degrees K., however, the universe was so cool that the only particles whose existence depended on an equilibrium with photons were pairs of positrons and electrons and pairs of neutrinos and antineutrinos. At that stage some protons and neutrons were left, but all the other heavy particles and antiparticles had been annihilated.

Before the temperature had fallen to 10^{11} degrees K. the density of matter, although still great by present standards, was low enough for the neutrinos and antineutrinos to cease interacting with the other particles. From that time on these neutrinos and antineutrinos went their separate ways. They are presumably still around, but at present there does not seem to be any means of detecting them.

When the temperature had dropped to 10^9 degrees K., the photons did not have enough energy to supply the rest-mass energy equivalent to the mass of the positron-electron pairs (according to the formula $E = mc^2$). Thus the equilibrium between the photons and the pairs of particles was disrupted: positrons and electrons that recombined to produce photons were no longer replaced by the reverse reaction. The positrons were steadily annihilated, leaving a small excess of electrons.

At about the same temperature the protons and neutrons underwent a series of nuclear reactions that resulted in the formation of helium nuclei (composed of either two protons and one neutron or two protons and two neutrons). Most of the helium in the universe today was formed at that time; the number of helium atoms observed in the present universe is an important check on conditions at that stage of the evolution of the universe. No free neutrons survived this stage. The universe at that time was composed of protons and helium nuclei that, together with the electrons, made up an ionized gas in thermal equilibrium with the photons. The enigmatic neutrinos were also still around. Photons in existence at this stage soon interacted with the ionized gas and do not survive to the present day.

When the temperature had dropped below about 5,000 degrees K., the nuclei and the electrons of the ionized gas recombined to give rise to an un-ionized gas. This gas did not interact further with the photons, so that those photons were undisturbed from that time on. They are what we detect when we observe the microwave background today. Since the time when they were created the Doppler shift caused by the expansion of the universe has weakened the radiation; although it started off at the time of recombination as black-body radiation with a temperature of some 5,000 degrees K. (corresponding to the radiation field at the surface of the sun), it is now black-body radiation at a temperature of only 2.76 degrees K.

The canonical hot-big-bang theory is the framework within which many astrophysicists are currently attempting to understand all the events in the evolution of the universe. In spite of the great leap in understanding brought about by the discovery of the microwave background, many questions remain to be answered, and the answers will need to be fitted into place. Why and how did galaxies form? Why are they the size they are and not larger or smaller? Why do they rotate? Where do their magnetic fields come from? Why and how did quasars form? Why are there fewer quasars now than there were in the past? These questions and many others are all being attacked by trying to find explanatory processes in the universe described by the canonical hot-big-bang model.

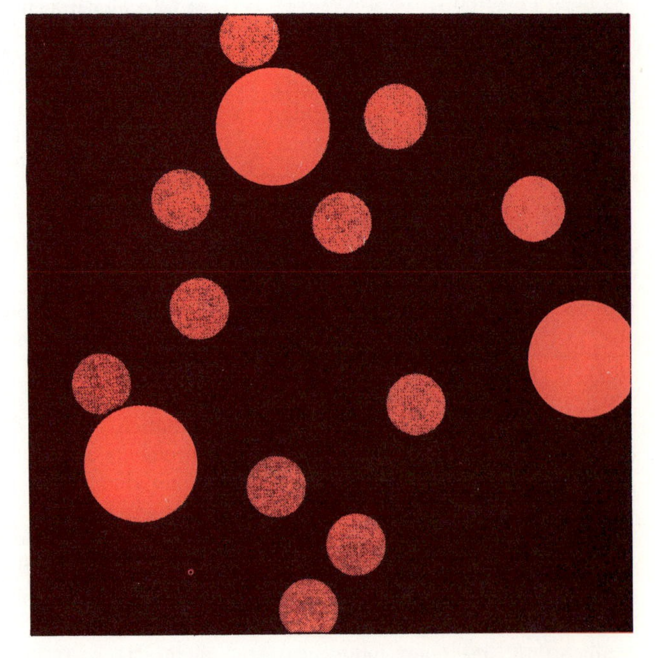

OLBERS' PARADOX states that if the universe were infinite in extent and all the stars were stationary with respect to one another, then the surfaces of the stars could be seen in all directions and the night sky would be very bright (*left*). Since such a phenomenon is not observed, something is wrong with that model of the universe. In fact, it is the expansion of the universe that both reddens and weakens the light of distant stars (*right*) so that the night sky is mostly black. Paradox also applies to other regions of spectrum.

There are some interesting problems at a more fundamental level that seem not so much to invite an answer from within the framework of the model as to demand an explanation in order that the model itself seem less arbitrary. Choosing to work backward from the present state of the universe to gain some knowledge of the initial conditions is not at all arbitrary, but it does not suffice to *explain* the initial conditions. Probably the most we can expect from this approach is that we shall be able at least to *describe* those conditions.

One unanswered question in this category is: Why is the universe expanding at all? We know it is expanding now, and as we work backward we calculate that it was expanding in all the earlier stages as well. But what started the expansion in the first place? There are other such questions, perhaps less obvious but no less important. For example, after the vast numbers of particles and antiparticles had annihilated themselves, why was there a small residuum of real particles (the protons, electrons and neutrons) that constitute the matter in the universe at the present time? Presumably there could just as easily have been a larger or smaller residuum of real particles, or a large or small residuum of antiparticles, or an exact cancellation of the two with no particles or antiparticles left at all. There is nothing in physics that would lead us to prefer any one of these possibilities. Why, then, is there precisely the observed density of real particles? This question goes right back to the initial conditions, because the most likely possibility is that the excess was always there and that it has been a characteristic feature of the universe at all times. It is not yet clear where the answers to the deeper questions are to be found, but these problems are certainly among the most challenging and stimulating in modern astronomy.

Let me return now to the observational situation. With the microwave background, as with the radio background, many attempts have been made to find differences in brightness between different directions in the sky, but no such differences are clearly apparent. The present state of the measurements is that variations in brightness in any angular area of the sky from 180 degrees across to about .1 degree across must be no greater than .1 percent. In other words, the microwave background is very uniform indeed.

It is possible to infer many interesting consequences from this fact. First, the solar system must be moving no faster

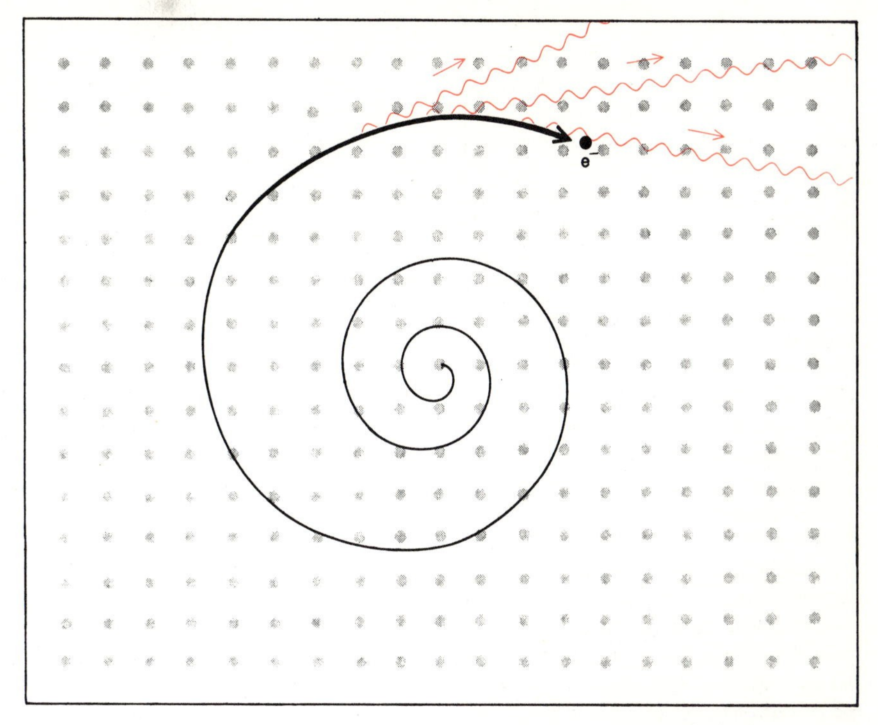

SYNCHROTRON RADIATION (*waves in color*) emitted by a charged particle (in this case an electron) spiraling in a magnetic field probably accounts for background radiation in radio region. Here the lines of magnetic force are perpendicular to the page (*gray dots*). Electron is spiraling up from the page and is emitting radiation tangent to its path.

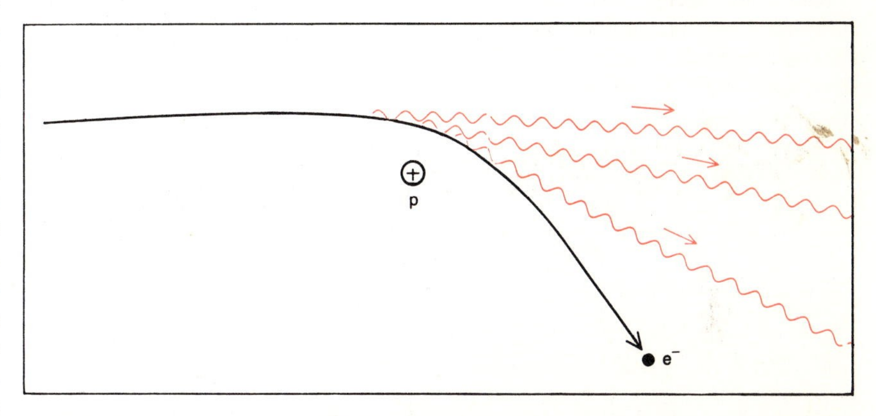

BREMSSTRAHLUNG RADIATION is generated when an electron passes so close to a nucleus (*plus sign in circle*) that its trajectory is bent. The slowing of the electron results in the emission of photons. This mechanism probably gives rise to the cosmic background radiation in the microwave region. (*Bremsstrahlung* is the German for braking radiation.)

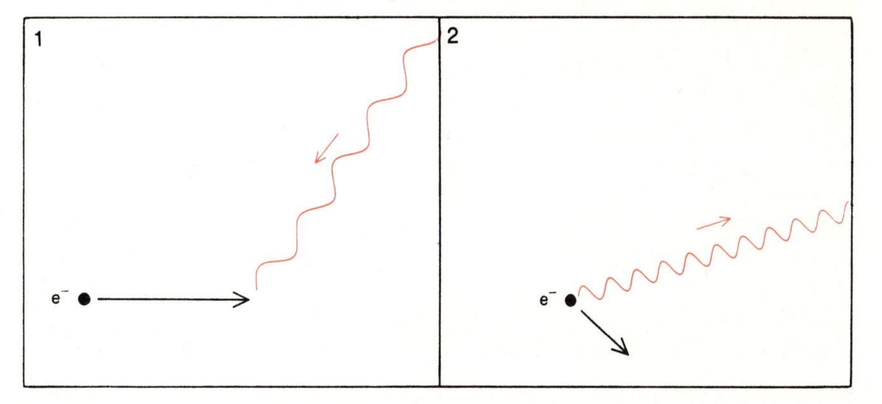

INVERSE COMPTON SCATTERING probably accounts for the background radiation in the X-ray and gamma-ray region. Here a low-energy photon impinges on a high-energy electron (*1*). Electron imparts some of its energy to photon and is itself slowed down (*2*).

than some 300 kilometers per second with respect to the frame of reference defined by the microwave background. If it were, the Doppler effect would cause the radiation from the direction in which the solar system is heading to be brighter than that from the rest of the sky by about .1 percent. It is likely that a small improvement in these observations will reveal the velocity of 200 kilometers per second of the solar system's rotation around the center of our galaxy.

A second consequence is that the universe must be expanding at the same rate in all directions. Otherwise the differences among the Doppler shifts of the radiation from the various regions of the recombining ionized gas that is responsible for the microwave background would cause brightness variations across the sky. A third consequence is that the universe cannot be rotating with any appreciable angular velocity or again the relativistic Doppler effect would give rise to observable brightness variations. The upper limit set for any possible rotation by these measurements is the phenomenally low value of a billionth of a second of arc per year.

The measurements of the uniformity of the background radiation on the smallest angular scales are designed to detect fluctuations in the density of the recombining ionized gas that condensed into the clusters and superclusters of galaxies in the present-day universe. Most calculations predict that the expected nonuniformities in brightness will be very small. Thus it is not too surprising that they have not yet been found.

An intriguing phenomenon within our own galaxy has been discovered through the microwave background. In the interior of certain small, dense clouds of gas and dust along the Milky Way there is a substantial concentration of formaldehyde molecules (H_2CO). Radio observations of these clouds have revealed an absorption line in their spectra at a wavelength of six centimeters, a wavelength characteristic of the formaldehyde molecule. At that wavelength the only continuous-spectrum radiation strong enough to be absorbed by the formaldehyde is the microwave background. The fact that the formaldehyde is absorbing energy rather than radiating it indicates that its temperature must be lower than the 2.76 degrees K. of the microwave background. In fact, the temperature of the formaldehyde is about one degree K. This raises the question of how the formaldehyde ever got so cold and how it stays that way. After all, if the formaldehyde were left to itself, it would be warmed to 2.76 degrees by the microwave background radiation. Some kind of previously unsuspected cosmic refrigerator must be working inside the clouds to keep the formaldehyde chilled.

In the X-ray region the cosmic background is represented by radiation at frequencies higher than 2.5×10^{17} Hz, corresponding to a wavelength of 1.3 nanometers. This radiation has a nonthermal spectrum that extends well into the gamma-ray region; the highest-energy photons detected so far have an energy of a little more than 100 million electron volts. The origin of the X-ray and gamma-ray background is not yet settled, but it certainly comes from outside our galaxy. The intensity of the radiation is much the same in all directions away from the plane of the Milky Way, where the sources would presumably be concentrated if they were within the galaxy.

The X-ray background is probably the sum of the radiation from a number of discrete sources, just as the radio background is the sum of the radiation from the radio galaxies and quasars. Attempts have been made to detect a graininess in the X-ray background that would reveal whether or not it is coming from a number of discrete sources, but the sensitivity of X-ray telescopes is not yet high enough to reveal the expected fluctuations. Moreover, too few extragalactic X-ray sources have been detected for the contribution of such sources to the X-ray background to be reliably calculated. Nonetheless, it is quite likely that the total emission from all X-ray sources can account for the entire background.

In the gamma-ray region the situation is not so clear. Gamma-ray telescopes are not yet sensitive enough to find any individual sources. The smooth continuity of the spectrum of the background radiation at the X-ray wavelengths and the gamma-ray wavelengths suggests that the gamma rays have the same origin as the X rays, but this need not be true.

The study of the background radiation in the X-ray and gamma-ray regions of the spectrum is still in its infancy. There is surely much interesting information to be gained from such investigations. The full value of even the present observations cannot be realized, however, until the basic question of the radiation's origin is settled. There is no shortage of possible mechanisms for generating the photons; the inverse Compton effect and the *Bremsstrahlung* mechanism are both likely candidates [*see illustrations on page 39*]. Nor is there any lack of hypotheses on the radiation's place of origin. The definitive observa-

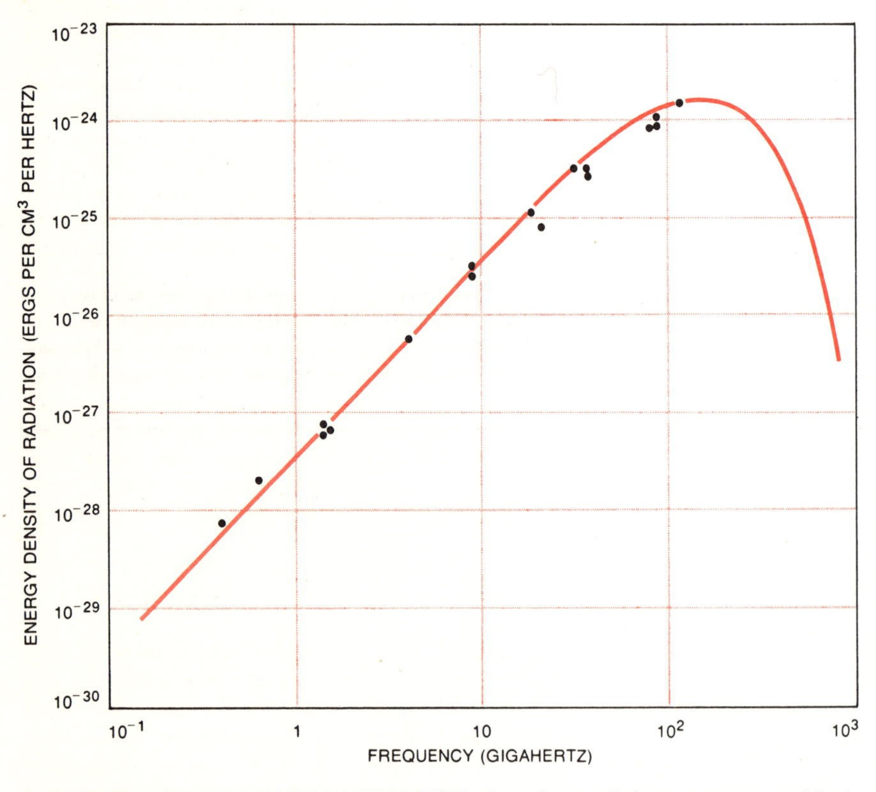

SPECTRUM OF MICROWAVE BACKGROUND shows how well the measurements (*dots*) fit the theoretical curve of a black body (an ideal radiator) at a temperature of 2.76 degrees Kelvin. Further measurements are needed in range of frequencies between 10^2 and 10^3 gigahertz (10^9 cycles per second) to confirm that the spectrum is indeed a black-body one.

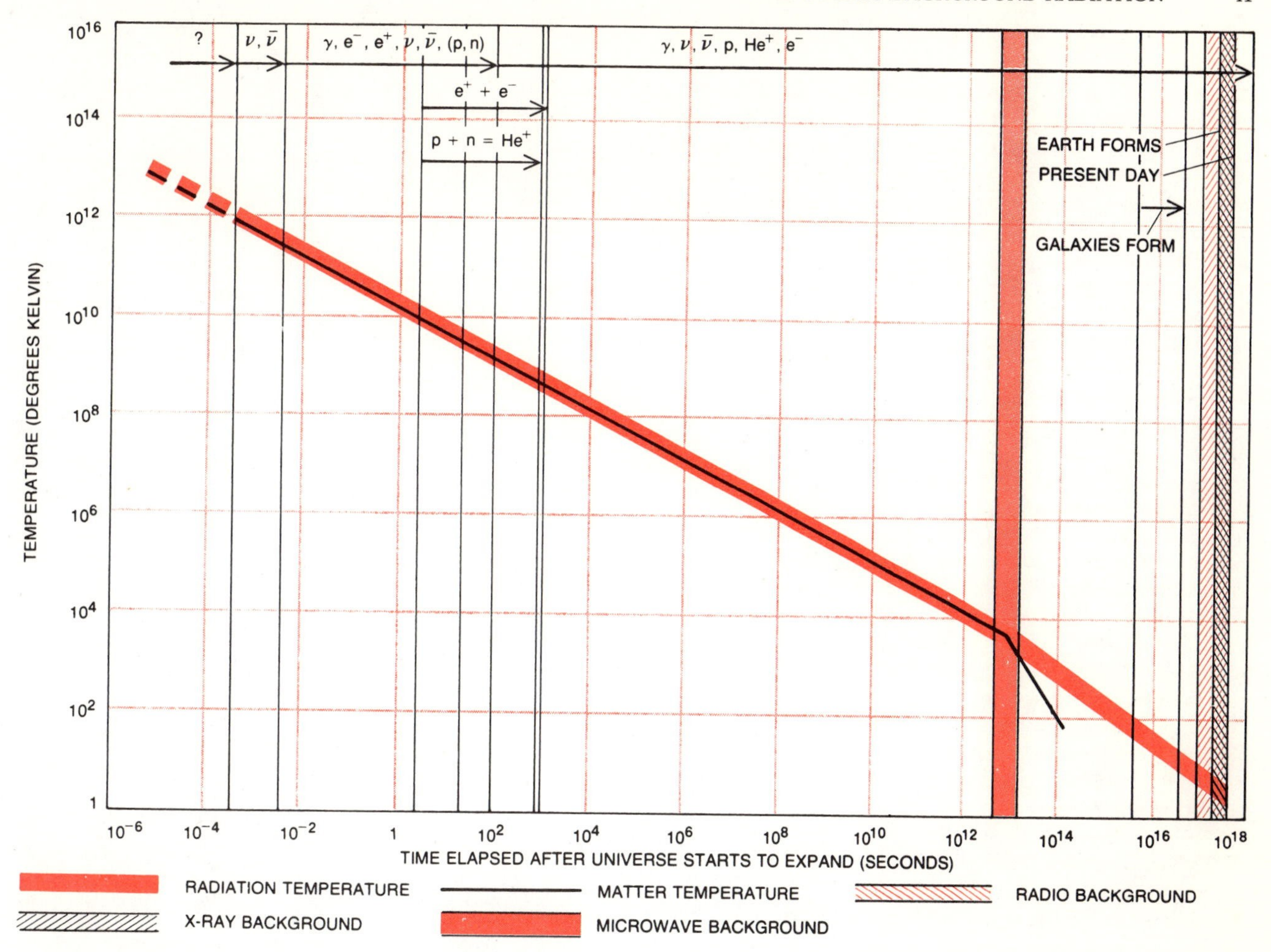

OUTLINE OF EVENTS IN THE UNIVERSE since the "big bang" shows when and how the cosmic background radiation originated in the various regions of the spectrum. Events in the universe within the first one-thousandth (10^{-3}) of a second after the big bang are not well understood, principally because they are dominated by interactions of nuclear particles that are not well understood. Up until a hundredth (10^{-2}) of a second after the big-bang neutrinos (ν) and antineutrinos (ν) are plentiful; they easily interact with photons (γ) and other particles. Thereafter the neutrinos and antineutrinos do not interact further, and so they play little part in subsequent physical processes. From 10^{-2} second until 100 seconds after the big bang the universe consists mostly of photons, electrons (e^-), positrons (e^+), neutrinos, antineutrinos and a trace of protons (p) and neutrons (n). During this time and continuing somewhat thereafter all the positrons combine with electrons ($e^+ + e^-$); in addition neutrons combine with protons to make helium nuclei (He^+). Most of the helium in the present universe was synthesized at this time. From 100 seconds after the big bang right up to the present the universe consists mostly of photons, neutrinos, antineutrinos, protons, helium nuclei and just enough electrons to keep everything electrically neutral. The particles of the ionized gas of hydrogen and helium interact frequently with the photons of radiation, so that the matter and the radiation stay at the same temperature. That temperature, however, is steadily decreasing as the universe expands. Somewhere in the interval between about 100 and 10^{14} seconds after the big bang the energy density of the electromagnetic radiation drops below the rest-mass energy of the matter, so that the photons and the particles no longer interact so freely. The large-scale dynamics of the universe change as a result, and the temperature drops off a little faster than before. At about 10^{13} seconds after the big bang electrons recombine with the ionized gas, emitting visible radiation that was subsequently Doppler-shifted by the expansion of the universe to the microwave region of the spectrum; the microwave background is this radiation from the recombination. The gas (matter) and the radiation now cool separately, each at its own pace; the final temperature of the radiation at the present time is 2.76 degrees K. Both the X-ray background and the radio background originate later, when X-ray sources and radio sources came into being. Too little is known about origin of gamma-ray background to include it in illustration.

tions have simply not yet been made.

In the infrared, visible and ultraviolet regions of the spectrum the cosmic background radiation has not yet been detected. There are technical difficulties in the way of some observations in these regions. For example, infrared and ultraviolet observations cannot be made from the ground because the earth's atmosphere strongly absorbs such wavelengths, and a balloon or a rocket is an exceedingly tricky observing platform. There are natural obstacles in the way of other observations. The faint glow of the earth's atmosphere at night, the zodiacal light and the faint stars in our galaxy are together at least 100 times brighter than the background radiation would be at the visible wavelengths. Direct observations of the background radiation may never be made in the far-ultraviolet region: neutral hydrogen atoms in the interstellar space of our galaxy absorb these photons so strongly that the extragalactic background radiation probably cannot reach the solar system at all.

As I have indicated, however, the extension of observations of the cosmic background radiation in other regions of the spectrum is far from hopeless. Indeed, it seems certain that we can look forward to observations that will lead to new and exciting inferences about the nature and history of the universe in the large.

The Evolution of Quasars

6

by Maarten Schmidt and Francis Bello
May 1971

It seems that they were much more plentiful when the universe was young than they are today. Their light takes so long to reach us that we can observe millions of them that have long been extinct

Since light has a finite velocity the astronomer can never hope to see the universe as it actually exists today. Far from being a handicap, however, the finite velocity of light enables him to peer back in time as far as his instruments and ingenuity can carry him. If he can correctly interpret the complex messages coded in electromagnetic radiation of various wavelengths, he may be able to piece together the evolution of the universe back virtually to the moment of creation. According to prevailing theory, that moment was some 10 billion years ago, when the total mass of the universe exploded out of a small volume, giving rise to the myriad of galaxies, radio galaxies and quasars (starlike objects more luminous than galaxies) whose existence has been slowly revealed during the past half-century.

Optical observations have shed little light on the evolution of ordinary galaxies because even with the most powerful optical telescopes such galaxies cannot be studied in detail if they are much farther away than one or two billion light-years. The astronomer sees them as they looked one or two billion years ago, when they were already perhaps eight or nine billion years old. Quasars, on the other hand, provide a direct glimpse of the universe as it existed eight or nine billion years ago, only one or two billion years after the "big bang" that presumably started it all.

Some 50 years ago the first large telescopes had shown that the light from distant galaxies is shifted toward the red end of the spectrum; the more distant the galaxy, the greater its red shift and the higher its velocity of recession. Like raisins in an expanding cosmic pudding, all the galaxies are receding from one another. From the observed velocities of recession one can compute that some 10 billion years ago all the matter in the universe was jammed into a tiny volume of space.

The term quasar, a contraction of "quasi-stellar radio source," was originally applied only to the starlike counterparts of certain strong radio sources whose optical spectra exhibit red shifts much larger than those of galaxies. Before long, however, a class of quasi-stellar objects was discovered with large red shifts that have little or no emission at radio wavelengths. "Quasar" is now commonly applied to starlike objects with large red shifts regardless of their radio emissivity.

This article is based on the hypothesis that the quasar red shifts are cosmological, that is, they are a consequence of the expansion of the universe and thus directly related to the distance of the object. On that hypothesis quasars are very remote objects. According to a contrary hypothesis, which will be discussed toward the end of the article, quasars are relatively close objects.

A recent study of quasars carried out with the aid of the 200-inch Hale telescope on Palomar Mountain has provided evidence that these extremely luminous objects evolved quite rapidly when the universe was young. The study indicates that quasars were about 100 times more plentiful when the sun and the earth were formed some five billion years ago than they are today. They were perhaps more than 1,000 times more plentiful at a still earlier epoch, say eight or nine billion years ago. Earlier than that, however, there may have been fewer quasars, perhaps because conditions in the universe had not yet favored their development.

The study embraced all the quasars in two areas of the sky representing a thousandth of the total celestial sphere. By extrapolation one can say with reasonable confidence that a complete sky survey with the largest telescopes should reveal on the order of 15 million quasars. The overwhelming majority are so far away that they almost certainly burned themselves out in the billions of years required for their light to reach us. All of them, of course, can still be studied telescopically, given the time and the inclination. If, however, it were possible to conduct an instantaneous survey of the universe, one might find that only about 35,000 quasars are in existence and radiating with their characteristic intensity at the present time. These find-

BRIGHTEST QUASAR and first member of its class to be recognized is 3C 273, indicated by the reticle in the photograph on the opposite page. The negative print was made from a 1-by-1⅜-inch portion near the edge of a 14-inch square plate taken with the 48-inch Schmidt telescope on Palomar Mountain as part of the National Geographic Society–Palomar Sky Survey. In 1962 the starlike object was found to coincide with the position of a strong radio source designated No. 273 in the third catalogue compiled by radio astronomers at the University of Cambridge. The optical magnitude of 3C 273 is 13. In the entire sky there are at least a million stars of that magnitude. A study of 3C 273's strange spectrum revealed, however, that its light was shifted toward the red end of the spectrum by an amount that indicated the object was receding at about one-sixth the velocity of light. This implied that it was not a nearby star but an object between one billion and two billion light-years away. A galaxy at the same distance would appear at least four magnitudes fainter, which means that 3C 273 is intrinsically at least 40 times brighter. The term "quasi-stellar radio source," or quasar, was coined to describe 3C 273 and other starlike objects exhibiting a large red shift.

ings are in conflict with the "steady state" hypothesis, which holds that the universe has always looked exactly the way it does today. That hypothesis postulates that new matter is continuously being created to maintain the expanding universe at a constant density.

After 10 years of intensive study by optical and radio astronomers quasars remain among the most puzzling of all celestial objects. Assuming that they are at cosmological distances, one can easily show that many quasars are from 50 to 100 times brighter than entire galaxies containing hundreds of billions of stars. Unlike the light output of normal galaxies, the light output of some quasars has been observed to change significantly in a matter of days. The only explanation is that some variable component of a quasar, if not the entire quasar, may be not much larger than the solar system.

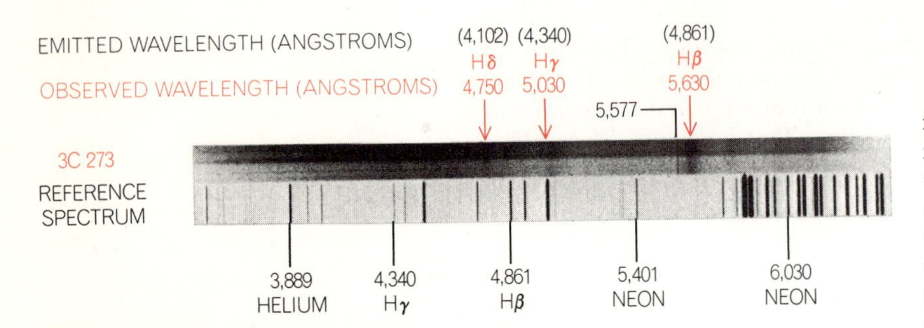

PORTION OF SPECTRUM OF QUASAR 3C 273 shows three prominent emission lines centered at 4,750, 5,030 and 5,630 angstroms, corresponding to the hydrogen emission lines delta, gamma and beta in the Balmer series. The upper and lower halves of the spectrogram were given different exposures to facilitate study. The three emission lines are produced by hydrogen atoms in various states of excitation. Two of the three lines, H gamma and H beta, also appear in the reference spectrum at their normal emission wavelength: 4,340 and 4,861 angstroms. The normal wavelength for H delta is 4,102 angstroms. The red shift, z, is obtained by subtracting the normal wavelength from the observed wavelength and dividing the difference by the normal wavelength. For 3C 273 z is .158, indicating the quasar is receding at nearly a sixth the speed of light. The sharp line at 5,577 often appears in spectra of astronomical objects and serves as a convenient reference point; it is produced by excited oxygen atoms in the upper atmosphere. This spectrogram and others in this article were made by one of the authors (Schmidt), who provided the original interpretation of 3C 273's spectrum.

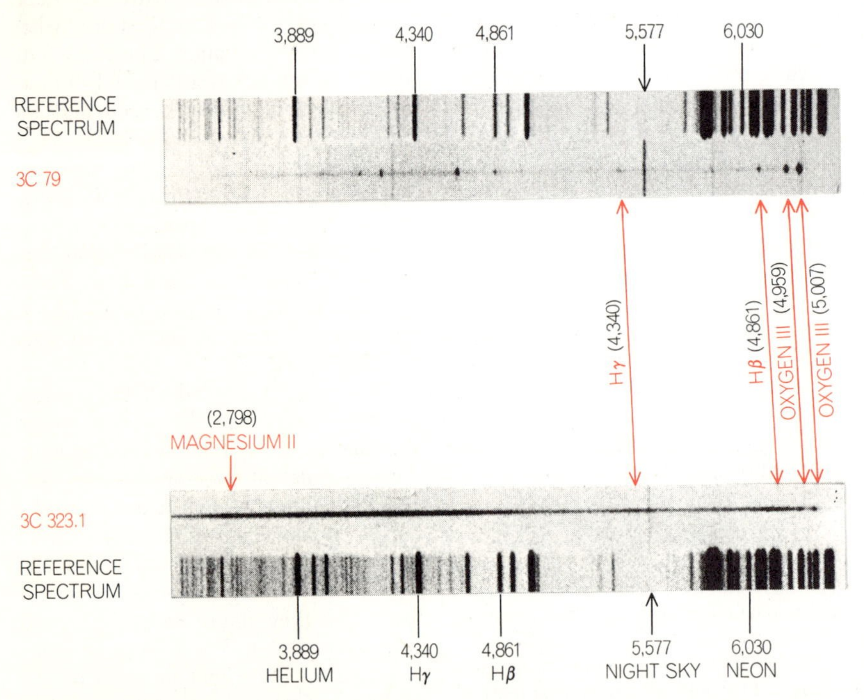

CONTRAST BETWEEN QUASAR AND RADIO GALAXY is shown by these two spectra. The spectrum at the top is that of the strong radio galaxy 3C 79, which has a red shift of .256. The bottom spectrum is that of quasar 3C 323.1, whose red shift is just slightly greater: .264. The radio galaxy produces a substantial number of sharp emission lines. Four of the lines in the right half of its spectrum are identified and compared with their much broadened counterparts as they appear in the spectrum of the quasar. Quasars characteristically emit strongly in the ultraviolet part of the spectrum. A common emitting ion is singly ionized magnesium, designated magnesium II, which has an emission wavelength of 2,798 angstroms.

The Discovery of Quasars

Before 1960 radio astronomers had identified and catalogued hundreds of radio sources: invisible objects in the universe that emit radiation at radio frequencies. From time to time optical astronomers would succeed in identifying an object—usually a galaxy—whose position coincided with that of the radio source. Thereafter the object was called a radio galaxy. The large majority of radio sources remained unidentified, however, and the general belief was that the source of the emission was a galaxy too far away, or at least too faint, to be recorded on a photographic plate.

In 1960 Thomas Matthews and Allan Sandage first discovered a starlike object at the position given for a radio source in the Third Cambridge ("3C") Catalogue, compiled by Martin Ryle and his colleagues at the University of Cambridge. The radio object 3C 48 coincided in position with a 16th-magnitude star whose spectrum exhibited broad emission lines that could not be identified. Not only did the object emit much more ultraviolet radiation than an ordinary star of the same magnitude but also its brightness varied by more than 40 percent in a year.

Object 3C 48 was thought to be a unique kind of radio-emitting star in our own galaxy until 1963, when the strong radio source 3C 273 was identified with a starlike object of 13th magnitude and one of the authors (Schmidt) recognized that most of the puzzling lines in its spectrum could be explained as the Balmer series of hydrogen lines, shifted in wavelength toward the red by 15.8 percent, or .158 [*see illustration on page 43 and upper illustration at left*]. Red shifts are commonly expressed as a fraction or percentage obtained by dividing the measured displacement of a line by the wavelength of the undisplaced line. With this clue it was immediately evident that the lines in the spectrum of 3C 48 had a red shift of .367 [see "Quasi-stellar Radio Sources," by Jesse L. Greenstein; SCIENTIFIC AMERICAN, December, 1963].

Such large red shifts, equivalent to a significant fraction of the velocity of

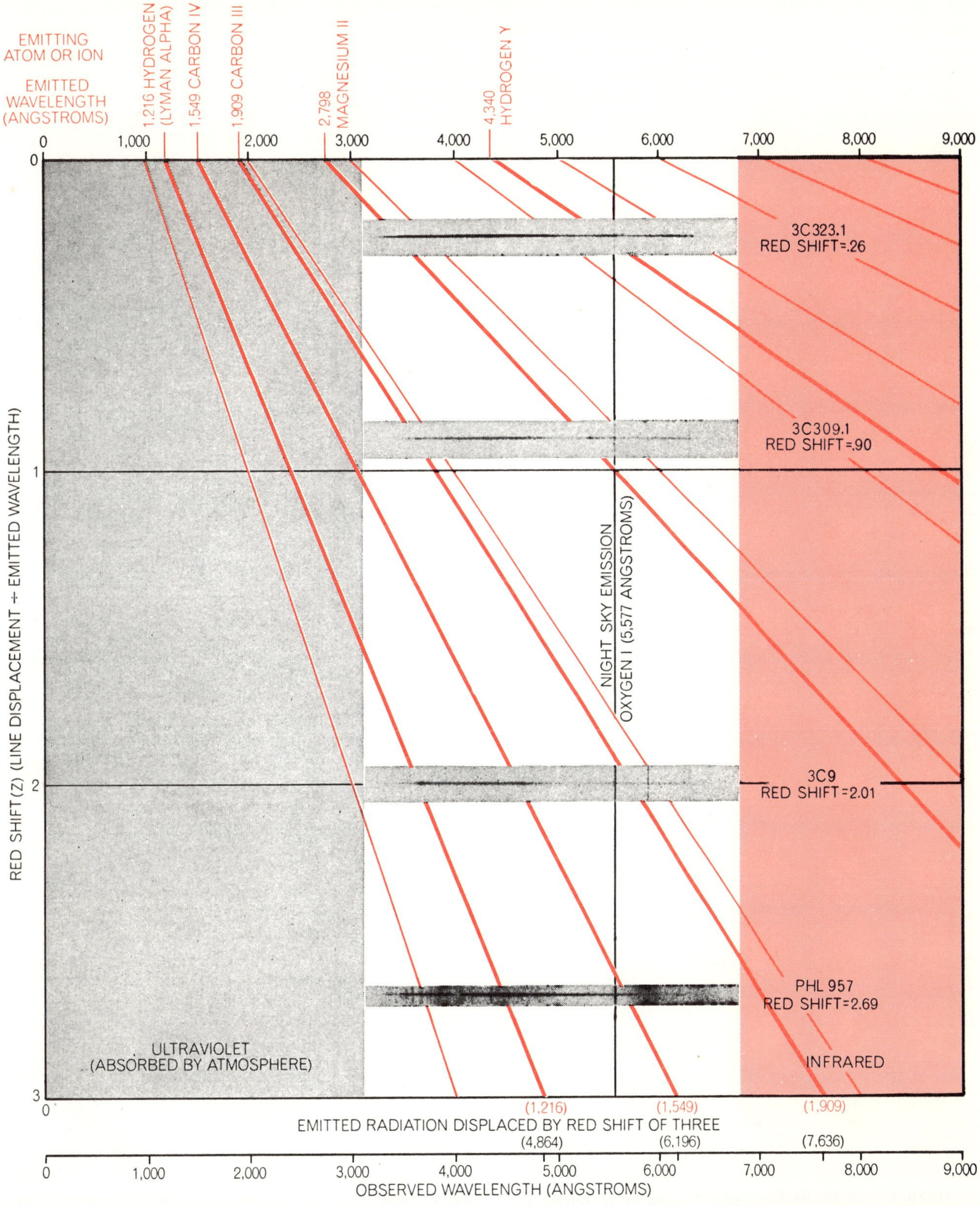

FOUR QUASAR SPECTRA are positioned on a diagram that shows how radiation emitted at one wavelength billions of years ago is "stretched" on its long journey through space by the presumed expansion of the universe. At least two emission lines are needed to establish the red shift of an astronomical object. A single line could represent any line shifted by any arbitrary amount. Here the heavy slanting lines correspond to the radiation emitted by hydrogen (Lyman alpha), carbon IV, carbon III, magnesium II and hydrogen (gamma). The roman numerals are one greater than the number of electrons missing from the atom. At a red shift, z, of 1 the Lyman-alpha line is observed at 2.432 angstroms; at a red shift of 2 the line is observed at 3,648 angstroms; at a red shift of 3 it would appear at 4,864 angstroms. Thus when z equals 2 the initial wavelength is stretched exactly three times; when z equals 3, four times and so on. The quantity $1 + z$ expresses how much the universe has expanded between the emission of a photon and its observation. Only two quasars are known with a red shift greater than 2.5; one of them is PHL 957, whose spectrum appears here. Its spectrum was made with an image-tube spectrograph; the other three spectra were recorded directly on photographic film. The photons that produced the spectrum of PHL 957 left the quasar when the universe was only about 13 percent of its present age.

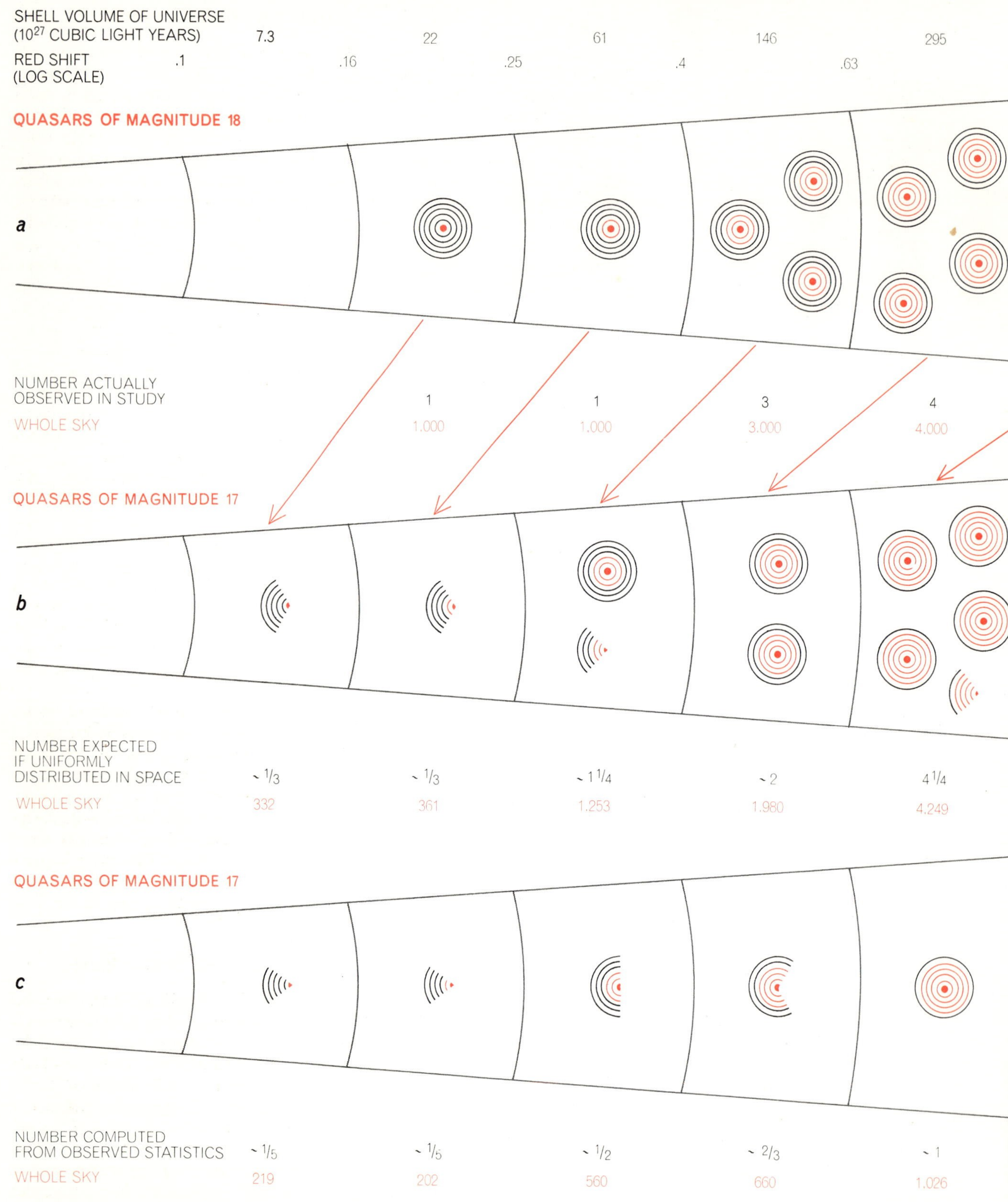

NUMBER OF QUASARS has been estimated by identifying and determining the red shifts of all the quasars in sample fields representing one-thousandth of the whole sky. The sample consisted of 20 quasars with an optical, or apparent, magnitude of about 18. It was clear from their red shifts, however, that some are much farther away than others and therefore are intrinsically brighter, as depicted in *a*. The red-shift intervals have been chosen so that the quasars in any given "shell" of the universe are on the average one magnitude (2.5 times) brighter in absolute luminosity than those in the next shell inward. Thus the four quasars in the sample box representing the most remote shell (red shift: 1.58 to 2.51) are each 100 times more luminous than the single quasar in the box whose red shift is between .16 and .25. Now, if quasars are uniformly distributed in space, and if there are 4,000 of maximum luminosity in the most remote shell, one would expect to find a proportional number in the next shell inward, whose volume is only two-thirds that of the outer shell. Two-thirds times 4,000 is 2,667. Thus diagram *b* shows that in the red-shift interval between 1 and 1.58 one would expect to find 2,667 quasars of maximum luminosity in the whole sky (or, proportionately, 2⅔ quasars in the small area actually sampled). In photographs these 2,667 should appear one magnitude brighter (magnitude 17) than the 4,000 of the same intrinsic brightness that are farther away. Using the same assumptions, one can estimate the number of quasars in still nearer shells.

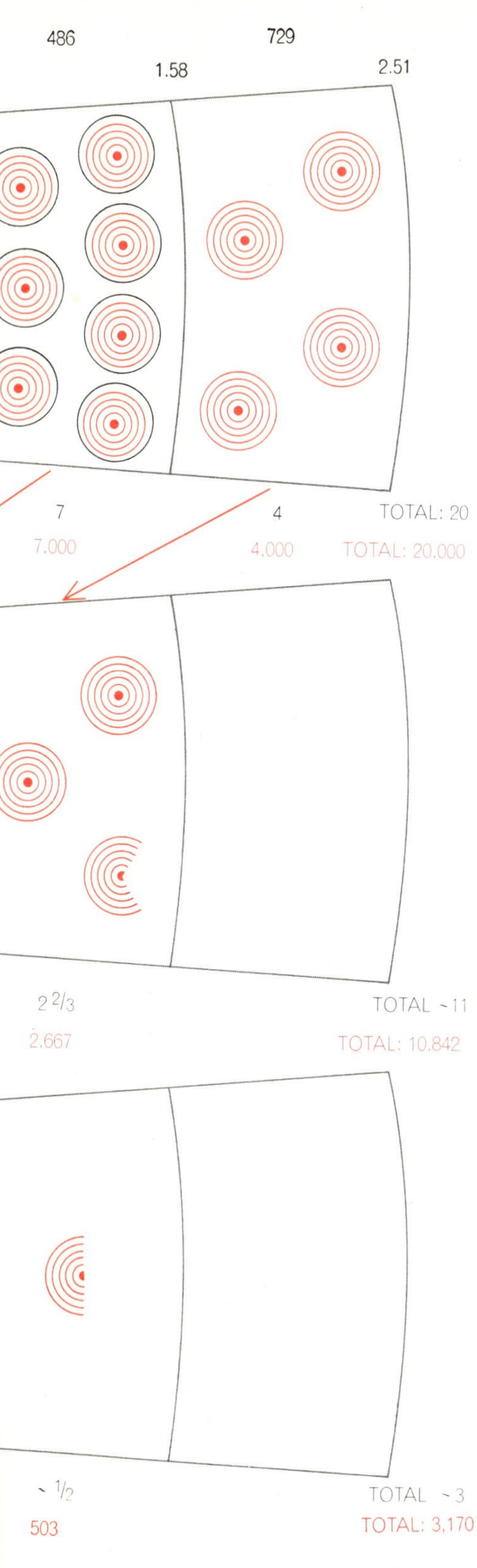

The total number is 10.842, distributed as shown in *b*. One concludes, therefore, that if quasars are uniformly distributed in space, one should observe about twice as many 18th-magnitude quasars as 17th-magnitude quasars. In actuality, however, surveys show that the number of quasars goes up by a factor of about six per magnitude. To satisfy this observation there can be only about 3,000 quasars of the 17th magnitude in the whole sky. An appropriate red-shift distribution for that approximate number is shown in *c*.

light, ruled out the possibility that 3C 273 and 3C 48 were stars in our galaxy. It was proposed that the red shifts are cosmological, which implies that the two objects have to be billions of light-years away and therefore extremely luminous to look as bright as they do in our night sky. They soon became known as quasars. Within the next few years quasars with even larger red shifts were discovered, including some with red shifts of more than 2, or more than six times the largest red shift ever observed for an ordinary galaxy. On the cosmological hypothesis, a red shift of 2 suggests that the light from the object has been traveling for about 80 percent of the age of the universe.

The Quiet Quasars

Several hundred radio sources have now been identified with starlike objects. Most of the identifications are made on the basis of positions provided by two or more radio telescopes spaced from several hundred meters to several thousand kilometers apart, used as an interferometer. The technique yields a precise measure of the difference in the time required for radio waves from the source to reach each telescope of the interferometer. One can then locate the source with an accuracy of between one second and 15 seconds of arc. Once the search has been narrowed to an optical candidate the final test is to see if its spectrum shows a red shift. More than half of the objects identified on the basis of their radio position usually turn out to be quasars. The spectroscopic test is unambiguous because the maximum red shift ever recorded for a star is .002; the smallest red shift for a quasar identified on the basis of its radio emission is .158 (for 3C 273).

It was noticed early that quasars usually emit rather strongly in the ultraviolet part of the spectrum. In 1964, when radio positions were known with considerably less accuracy than they are today, Ryle and Sandage conceived the idea of using ultraviolet strength as a clue in searching for optical counterparts of radio sources. They used a technique in which a single photographic plate of a star field was exposed to blue light and then was shifted slightly and exposed to ultraviolet radiation. By visual examination it was possible to readily distinguish strong ultraviolet emitters from normal stars.

In 1965 Sandage noted that objects with excess ultraviolet emission were much more plentiful than known radio sources in typical star fields. He soon discovered that some of these "blue stellar objects" exhibit red shifts that qualify them as quasars even though no radio emission has been detected from them. Most of the other strong ultraviolet emitters turn out to be white-dwarf stars in our own galaxy. Thus only a small fraction of quasars are strong radio emitters. The rest are radio quiet, or virtually so. It may be that a typical quasar is a strong radio emitter for only a small part of its life-span. Alternatively, it may be that relatively few quasars are born to be strong radio sources.

Two years ago Sandage and Willem J. Luyten published photometric analyses of 301 blue objects in seven survey fields. They counted quasar candidates tentatively selected from these blue objects and estimated that in one square degree of the sky (roughly equal to five times the area of the moon) there is, on the average, .4 quasar brighter than magnitude 18.1. They also estimated that there are five quasars per square degree brighter than magnitude 19.4, and that tentatively there are as many as 100 brighter than magnitude 21.4. Over the entire sky they estimated there may be 10 million quasars of the 22nd magnitude or brighter.

The greater the magnitude, of course, the dimmer the object; every increase of five magnitudes (say from the 18th magnitude to the 23rd) corresponds to a decrease by a factor of 100 in brightness. The number of quasars increases steeply with magnitude, by a factor of about six per magnitude. This steep increase is incompatible with a uniform distribution of quasars in space, as we shall see.

Counting Verified Quasars

The objects isolated by Sandage and Luyten are defined as "faint blue objects [with] an ultraviolet excess." For a detailed statistical study one has to obtain the spectrum of each "faint blue" candidate individually to establish whether or not it is really a quasar. One of the authors (Schmidt) began this task about four years ago, working with several of the star fields examined by Sandage and Luyten. The ultimate goal of the study is to establish how quasars are distributed by red shift (distance) and luminosity.

Of 55 faint blue objects investigated in two of the Sandage-Luyten fields, 32 turned out to have negligible red shifts and therefore could be rejected as being dwarf stars within our own galaxy. The 23 remaining objects exhibited spectra characteristic of quasars, and all but one of the spectra contained the minimum of two lines needed for establishing a red shift. A single line could represent almost any emitting atom red-shifted by any arbitrary amount. When a spectrum contains two lines, however, it is almost always possible to assign a unique red-shift value that identifies a reasonable emission wavelength for each line [*see illustration on page 45*]. Unfortunately the spectra of some quasars show only a single clear line, thereby frustrating efforts to establish their red shift. Although the red shifts assigned to several of the objects are still tentative, the overall distribution must be essentially correct. The red shifts range from .18 to 2.21. None of the 23 quasars appears in any of the catalogues of strong radio sources.

At this stage it will be most useful in our discussion to concentrate on the quasars of 18th magnitude in the sample. There are 20 such quasars. Since the 20 objects exhibit a variety of red shifts, however, we know they must lie at vastly different distances and therefore must differ greatly in *absolute* luminosity even though they look equally bright to an observer.

To express these differences in absolute luminosity one can classify the objects by red shift in such a way that each red-shift category represents a step of one magnitude in absolute luminosity. The relation between the red shift and the magnitude of a standard source depends on the properties of the universe. In the cosmological model followed in this study a quasar of 18th magnitude whose red shift falls in the range between .25 and .4 is intrinsically brighter by one magnitude than a nearer object whose red shift lies between .16 and .25. Six red-shift categories, each corresponding to a step of one magnitude in absolute luminosity, are enough to cover the range of red shifts actually exhibited by the 20 objects. The brightest members of the group are five magnitudes, or 100 times, brighter than the least luminous.

When the 20 quasars were grouped by red shift in this way, their distribution was found to be similar to the red-shift distribution of radio quasars of the same optical magnitude. Taking this into account and rounding things off somewhat, the following distribution for the red shifts of 18th-magnitude quasars was adopted for the subsequent analysis:

Red shift	1.58–2.51	20 percent
Red shift	1.00–1.58	35 percent
Red shift	.63–1.00	20 percent
Red shift	.40– .63	15 percent
Red shift	.25– .40	5 percent
Red shift	.16– .25	5 percent

The Sandage-Luyten survey had shown that in the entire sky there are, in round numbers, 20,000 quasars of apparent magnitude 18—just 1,000 times as many as in the new detailed sample.

REDSHIFT Z	SHELL VOLUME OF UNIVERSE (10^{27} CUBIC LIGHT YEARS)	NUMBER OF QUASARS IN WHOLE SKY: MAGNITUDE 17 CORRECTED DISTRIBUTION $\sim(1+Z)^6$	MAGNITUDE 17 IF UNIFORMLY DISTRIBUTED	MAGNITUDE 18 DERIVED FROM OBSERVATION
1.58–2.51	729			4,000
1.00–1.58	486	503	2,667	7,000
.63–1.00	295	1,026	4,249	4,000
.40– .63	146	660	1,980	3,000
.25– .40	61	560	1,253	1,000
.16– .25	22	202	361	1,000
.10– .16	7.3	219	332	
		3,170	10,842	20,000

DISTRIBUTION OF QUASARS according to red shift is shown for 20,000 quasars of 18th optical magnitude (*column at far right*), based on a representative sample of 20 quasars. The adjacent columns present two different estimates of the total number of quasars of the 17th magnitude. The method of making the estimates is explained in the illustration on the preceding two pages, where the same numbers appear in the diagrams labeled *b* and *c*. Observation shows that the number of quasars goes up by a factor of about six per magnitude rather than the factor of two expected if quasars were uniformly distributed throughout space. One can obtain the observed distribution by multiplying the shell volume of the universe by $(1+z)^6$, where z is the red shift and the exponent 6 is an experimentally determined value that yields the desired increment per magnitude. The table at top of these two pages shows the computed number and red shift of all quasars from magnitude 13 through 23.

REDSHIFT Z	APPROXIMATE VISUAL MAGNITUDE: SHELL VOLUME OF UNIVERSE (10^{27} CUBIC LIGHT YEARS)	SHELL VOLUME OF UNIVERSE $X(1+Z)^6$
1.58 –2.51	729	593,000
1.00 –1.58	486	74,600
.63 –1.00	295	10,900
.40 – .63	146	1,800
.25 – .40	61	336
.16 – .25	22	68
.10 – .16	7.3	15
.06 – .10	2.1	3.4
.04 – .06	.59	.8
.025– .04	.16	.19
.016– .025	.04	.05

TOTAL QUASAR POPULATION of universe is estimated to be on the order of 14 million, of which more than 99.7 percent are evidently fainter than the 18th magnitude and have red shifts greater than .4. From the 13th to 18th visual magnitude the number of quasars increases by a factor of five or six

If the 20,000 are distributed according to the percentages listed above, one finds that the number in each red-shift category, starting with the highest, is as follows: 4,000, 7,000, 4,000, 3,000, 1,000 and 1,000. It is clear that in a random sample of 18th-magnitude quasars more than half are extremely distant (red shift greater than 1) and therefore belong to the most luminous members of their class. A red shift of 1 corresponds to looking back two-thirds of the time that has elapsed since the universe began its expansion.

Proceeding to the next stage of the analysis, one would like to estimate the number of quasars whose apparent magnitude is either brighter or fainter than 18 and how they are distributed according to red shift. To do this one must know the volumes of the successive shells of the universe in which we have placed our 18th-magnitude quasars. These volumes depend on the cosmological model followed. Our unit of volume, 10^{27} cubic light-years, or a cube of which each side is a billion light-years, is co-moving, which means that no matter what the red shift, the unit of volume expands with the universe into our "local" unit of 10^{27} cubic light-years.

The 4,000 brightest and most distant quasars (red shift 1.58 to 2.51) occupy a shell with a volume of 729×10^{27} light-years. The problem now is to use this in-

APPROXIMATE VISUAL MAGNITUDE										
13	14	15	16	17	18	19	20	21	22	23
—	—	—	—	—	4,000	56,000	217,000	1,000,000	2,000,000	9,000,000
—	—	—	—	503	7,000	27,000	124,000	200,000	1,000,000	—
—	—	—	74	1,026	4,000	18,000	32,000	200,000	—	—
—	—	12	169	660	3,000	5,000	27,000	—	—	—
—	2	32	123	560	1,000	5,000	—	—	—	—
0	6	25	113	202	1,000	—	—	—	—	—
1	6	25	44	219	—	—	—	—	—	—
1	6	10	50	—	—	—	—	—	—	—
1	2	12	—	—	—	—	—	—	—	—
1	3	—	—	—	—	—	—	—	—	—
1	—	—	—	—	—	—	—	—	—	—
5	25	116	573	3,170	20,000	111,000	400,000	1,400,000	3,000,000	9,000,000

for each decline of one magnitude in brightness. Beyond the 18th magnitude, however, the increase is slower because the table contains no entries for quasars with red shifts greater than 2.51. In fact, only two quasars with larger red shifts are known, which suggests that there is a genuine paucity of such objects. Any quasar with a red shift of 2.5 is so distant that its light has been traveling through space for more than 85 percent of the age of the universe. The light from the more than 13.5 million quasars with a red shift greater than 1 has been en route for at least 6.8 billion years, assuming that the universe is on the order of 10 billion years old. Because the lifetime of a quasar is probably well under a billion years, the overwhelming majority of all the quasars that ever existed must have evolved by now into less luminous objects, perhaps ordinary galaxies. One can estimate that only about 35,000 quasars exist today.

formation to compute how many quasars of the same absolute luminosity would appear in the shell immediately within the outermost one, whose red shift corresponds to between 1 and 1.58. That shell, according to the cosmological model selected, has a volume of 486×10^{27} cubic light-years, or two-thirds of the volume of the outer shell. Now we introduce a supposition. If quasars were uniformly distributed in space, the inner shell would contain two-thirds times 4,000, or 2,667, quasars exactly like those in the outer shell. If quasars of that intrinsic luminosity were moved one shell closer to us, their apparent luminosity, as we observe them, would therefore be one magnitude brighter, that is, magnitude 17 instead of magnitude 18 [*see illustration on pages 46 and 47*]. Remember that the red-shift intervals were chosen specifically so that each step would correspond to a one-magnitude change in brightness.

A similar computation is now performed for the 18th-magnitude quasars in each of the other red-shift categories. In each case one computes the number expected in the shell within the preceding one, assuming as before that quasars are uniformly distributed in space.

This calculation yields the following additional numbers: 4,249, 1,980, 1,253, 361 and 332. When these are added to the number 2,667 previously computed, one obtains a total of 10,842 quasars of apparent magnitude 17, or roughly half as many quasars as one expects to find of magnitude 18 (assuming uniform distribution).

We recall that the Sandage-Luyten survey shows that the number of quasar-like objects increases not by a factor of two per magnitude (from 10,842 to 20,000 in the exercise just completed) but by a factor of about six. In other words, their statistics would predict only some 3,000 or 4,000 quasars of apparent magnitude 17 rather than 10,842.

What the factor of six tells us, of course, is that there are more faint quasars than one would expect to find if space were uniformly filled with quasars. The only plausible explanation is that the density of quasars must increase with increasing distance, that is, as we look back farther in time. To arrive at a distribution law that satisfies the observational evidence, let us assume that the density is proportional to some power, n, of the scale of the universe. The scale, or size, of the universe is inversely proportional to the amount by which light has been "stretched" by the expansion of the universe. Thus if the Lyman-alpha line emitted at 1,216 angstroms is observed at 3,648 angstroms, one can say that the universe has expanded by a factor of three since the radiation left the emitter. Since the red shift, z, in this case is 2 (3,648 minus 1,216 divided by 1,216) it is evident that the scale of the universe is given not by z but by $1 + z$. The density law we are seeking is therefore $(1 + z)^n$.

The value of n is simply obtained by trial and error to yield about 3,000 quasars of magnitude 17 [*see illustration on opposite page*]. Quite by accident the value of n turns out to be 6. It is only a coincidence that n is 6 and that the increase in the number of quasars per magnitude is also six. With this density law it is a simple matter to extend the distribution table downward from magnitude 17 and upward from magnitude 18 [*see illustration above*].

Along the bottom of the table one can read off the number of quasars expected in the entire sky for each magnitude. For the five magnitudes brighter than magnitude 18 the expected quasar population decreases steadily at each step from 3,170 (17th magnitude) to 573 (16th) to 116 (15th) to 25 (14th) and finally to five (13th). For the five magnitudes fainter than 18 the expected population rises steeply at each step from 111,000 (19th magnitude) to 400,000 (20th) to 1.4 million (21st) to three million (22nd) and finally to nine million (23rd). The total estimated quasar population from magnitude 13 to magnitude 23 inclusive is thus about 14 million.

The table does not list entries for

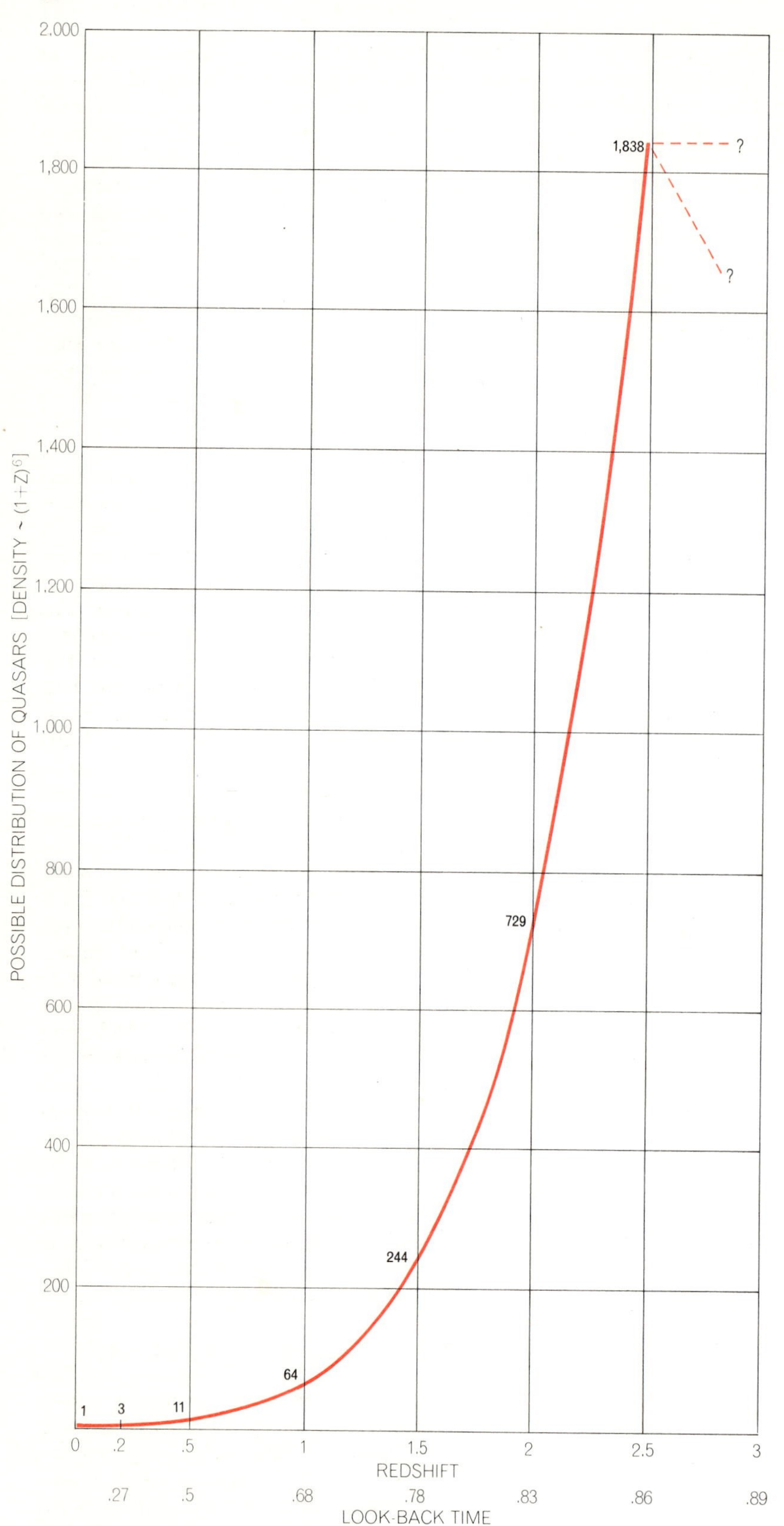

CHANGE IN QUASAR DENSITY WITH TIME can be derived from the table on the preceding two pages. The curve shows that the number of quasars rises steeply with increasing red shift, which is equivalent to looking back in time. Thus if one looks back 68 percent of the age of the universe, one would find more than 60 quasars in the volume of space that now contains one quasar. Looking back 83 percent of the age of the universe, one would find more than 700 quasars in the same volume. The maximum density may have existed when the universe had reached only about 14 percent of its present age. The scarcity of quasars with a red shift greater than 2.5 suggests that their density was no greater at earlier epochs.

quasars with red shifts greater than 2.5. Actually two quasars with larger red shifts are known: one, PHL 957, has a red shift of 2.69; the other, 4C 05.34, has a red shift of 2.88. Their magnitudes are respectively 17 and 18. If the density law $(1 + z)^6$ continued to hold, one would expect a great many 19th-magnitude quasars with red shifts larger than 2.5. Their scarcity suggests that the density does not increase beyond 2.5 and that it may actually decrease.

The probable scarcity of quasars with red shifts greater than 2.5 implies that the largest telescopes are able to look back in time to the epoch when quasars made their first appearance in the universe. Depending somewhat on the cosmological model selected, one can say that the light from a quasar with a red shift of 2.5 began its journey through space some 8.6 billion years ago, or some 1.5 billion years after the big bang that hypothetically created the universe as we know it. Within the next few billion years the great majority of quasars were born and began their brief but brilliant career [*see illustration at left*].

One can estimate that the universe at present contains only some 35,000 quasars. All the rest have presumably evolved into less remarkable objects, perhaps ordinary galaxies; we know of their existence because the signals they emitted billions of years ago are only now reaching our telescopes. The quasars of the lowest intrinsic luminosity (those at the bottom of the table on the preceding two pages) are no brighter than large galaxies. It is therefore uncertain whether all of them are quasars or whether some are compact galaxies of one kind or another. To avoid such confusion one could consider leaving out the quasars listed in the two lowest (least luminous) categories all across the table. The remaining "high luminosity" quasars would then number about 1.5 million for the entire sky, and the number existing at the present time would drop to only 3,500.

Another way to look at the quasar population developed in this analysis is to compare the number of quasars with the number of galaxies in a given volume of space. A volume of 10^{27} cubic light-years in our neighborhood contains about 20 quasars, of which two are objects of high luminosity. In very round numbers the same volume of space contains probably between one million and 10 million galaxies.

The study described above involved quasars selected solely on the basis of their optical properties; their radio emission, if any, is negligible. It is therefore important to ask if quasars selected on the basis of their radio luminosity also show an increase in density with increasing distance. The 3C catalogue mentioned above is a comprehensive listing of all radio sources in the northern half of the sky with a certain minimum radio intensity. (The minimum value is nine "flux units" at 178 megahertz, or 9×10^{-26} watt per square meter per hertz.) By the late 1960's 44 of the 300-odd extragalactic radio sources in the 3C catalogue had been optically identified as quasars. Of these 44 objects 33 had optical magnitudes of 18.5 or greater, and there was reason to believe that the 33 represented essentially all the 3C quasars down to that limiting magnitude.

Radio-bright Quasars

The analysis of the distribution of the 33 objects is complicated because both a radio limitation and an optical limitation were involved in their selection. That is to say, to appear in the group of 33 quasars an object had to radiate strongly in two widely separated parts of the spectrum: the radio region and the optical region. The analysis made by one of the authors (Schmidt) went as follows:

From the red shift the distance to each object was computed on the basis of some particular model of the expanding universe. This distance equaled the radius of the volume of space within which the object was actually observed. One can then ask how far the object could be moved outward before one of two things happen: either its apparent magnitude drops below 18.5 or its radio flux falls below nine units. This distance defines the radius of the maximum volume beyond which the object could not lie and still remain a member of its original class.

For each object one can express the ratio of the two volumes, actual volume over maximum volume, as a decimal fraction. A priori, if the 33 objects were uniformly distributed, one would expect the average value of this fraction to be .5. Thus one would expect half of the values to be less than .5 and the other half to lie between .5 and 1. Actually only six of the objects yield values below .5 whereas 27 give higher values. In other words, radio quasars tend to occupy the outer reaches of the volume within which they can be observed. This tells us that their density increases with distance. When the density law is worked out in detail, it is found to lie between $(1 + z)^5$ and $(1 + z)^6$. That is remarkably similar to the density law obtained for optically selected quasars, which on the average show negligible radio emission. The conclusion is that quasars have a density distribution that is only slightly or not at all dependent on their radio properties. This still leaves unsettled, however, the two possibilities already mentioned: either most quasars pass through a brief evolutionary stage during which they emit strongly at radio wavelengths or else only a small fraction of all quasars are destined to evolve into strong radio emitters.

Other Quasar Hypotheses

A number of astronomers and theorists originally found it difficult to accept the idea that the red shifts of quasars are cosmological. They did not see how it was possible for an object to emit as much light as 100 galaxies and yet vary in intensity by 10 percent or more in a few days. They proposed, as one alternative, that quasars might be much nearer and smaller objects ejected at high velocity from the center of our own galaxy. This is sometimes called the local-Doppler hypothesis because the red shift is a Doppler shift and the objects are of local origin. Being only a few million light-years away, rather than billions of light-years, their actual energy output would be much less.

This hypothesis has encountered the difficulty that quasars are much more numerous than anyone suspected in the early 1960's. As we have just seen, recent estimates run into the millions, and on the most conservative basis one can hardly assume fewer than a million quasars. It may be estimated that the mass of the typical quasar, on the basis of the local-Doppler hypothesis, would have to be at least 10,000 suns. The ejection of a million objects, each of 10,000 solar masses, from the center of our galaxy would require that the mass of all the stars in the galactic nucleus be completely converted into energy. One must also explain why the only quasars ever observed are those ejected by our own galaxy. If any quasar-like objects had been ejected by any of the scores of galaxies in our immediate neighborhood, some of them should be observed to be heading *toward* us and thus should exhibit a blue shift rather than a red shift. Yet no quasar-like object with a blue shift has ever been detected. The local-Doppler explanation, on the whole, must be regarded as being quite unlikely.

A totally different explanation for the red shift of quasars seemed attractive at first. According to this hypothesis quasars are objects in which a substantial mass is compressed into an extremely small volume. Light emitted from such an object would have to overcome an immense gravitational potential and would be red-shifted just as it is in quasars. The physical conditions that the hypothesis must account for can be rather precisely calculated. It is possible to compute, therefore, how large an emitting envelope of gas is needed, and what its density and temperature must be, to produce the spectral lines actually observed in quasars.

But if one assumes, to take an extreme case, that the highly condensed mass is comparable to the mass of the sun, its emitting envelope would not exhibit the required luminosity unless it were within 10 kilometers of the observer! The object has to be more distant, of course, and that will require a larger mass. The masses computed are large, and thus tend to create inadmissible side effects. For example, at a distance of 30,000 light-years the mass would have to be 10^{11} suns; it would rival the mass of our own galaxy, whose center is at the same distance. If the mass is raised still further to 2×10^{13} suns, the minimum distance can be raised to 10 million light-years. In that case, in order not to raise the observed average density of the universe, a million such quasars would have to be distributed out to a distance of at least a billion light-years, at which point they would hardly qualify any longer as local objects.

One other "anticosmological" hypothesis should be mentioned for the sake of completeness: the hypothesis that the cause of the quasar red shift is simply unknown and thus lies outside present-day physics. Since no arguments can be made against such a metaphysical hypothesis it cannot be excluded.

The Cosmological Hypothesis

An attractive feature of the cosmological hypothesis is that the quasar red shift comes "free," without requiring the introduction of bizarre physical conditions to explain the shift. The quasars exhibit a red shift simply because they are being carried along by the expansion of the universe. The extraordinary luminosity of quasars, together with their short-term variability, originally constituted the strongest objection to the cosmological hypothesis. In the past five years, however, short-term luminosity fluctuations of considerable magnitude have been observed in the nuclei of two rather special kinds of galaxy: N-type galaxies and Seyfert galaxies. These nuclei are starlike and resemble quasars in producing an excess of ultraviolet radiation. Moreover, there is general agreement that their red shifts, even though they are modest in the case of Seyfert galaxies, are cosmological in origin.

Recently it has been found that both quasars and Seyfert galaxies radiate strongly in the infrared region of the spectrum. Indeed, the infrared luminosity of the nearby Seyfert radio galaxy 3C 120 is 10^{46} ergs per second, which is equal to the infrared luminosity of many quasars when their luminosity is calculated on the assumption of their being at cosmological distances. In other words, we now have examples of objects whose extraordinary energy output is as difficult to explain as the output of quasars (regarded as cosmological objects) and whose output varies over time scales that are just as brief as the time scales for the variation of quasars. Therefore the cosmological hypothesis cannot be ruled out on the basis of the difficulties encountered in explaining the quasars' rapidly varying high luminosity, because the same difficulties hold for galaxies whose properties and distances are not in question.

Support for the cosmological hypothesis has recently been obtained by James E. Gunn of the Hale Observatories. He found that the image of the quasar PKS 2251 + 11 (red shift .323) is superposed on the image of a small, compact cluster of galaxies. Gunn was able to determine the red shift of the brightest galaxy in the cluster and found a value of $.33 \pm .01$. The coincidence in direction and red shift makes it very likely that the quasar is associated with the cluster of galaxies, thus confirming the cosmological nature of its red shift.

As for the ultimate source of the tremendous energy observed in quasars, there has been no lack of hypotheses, among them stellar collisions, the gravitational collapse of massive stars, supernova explosions, conversion of gravitational energy into particle energy by magnetic fields, matter-antimatter annihilation and the rotational energy of a very compact mass (as proposed for pulsars). There is also no agreement about the radiation mechanism, particularly in the infrared, where much of the output is radiated. Similar problems exist for nuclei of galaxies, notably for those of Seyfert galaxies. The solution of these problems constitutes one of the main challenges to present-day astronomy.

ears ago, and so we can be sure
ome galaxies are even older than
On the other hand, as we have
ned, all the galaxies must have
ightly packed together no more
0 billion years ago, based on their
t recession velocity. Estimates of
es of stars suggest that our galaxy,
thers like it, are unlikely to be
less than 10 billion years old.
we are presented with a remark-
oincidence: most galaxies appear
about as old as the universe. This
s that galaxies must have formed
conditions in the universe were
different from those now prevail-

ms clear, then, that the formation
galaxies cannot be treated apart
cosmological considerations. The
ics and structure of the universe
large are beyond the scope of New-
physics; it is necessary to use
in's general theory of relativity.
se of the complexity of the theory,
racticable to solve the equations
or cases having special symmetry.
quite recently the only solutions
expanding universe were those
in 1922 by the Russian mathema-
Alexander A. Friedmann. In his
ed models matter is treated as a
uniform and homogeneous me-
The universe expands from a sin-
state of infinite density, with the
expansion decelerating as a con-
nce of the mutual gravitational at-
n of its different parts. The uni-
may have enough energy to keep
ding indefinitely or the expansion
eventually cease and be followed
general collapse back to a com-
d state. Observations of the actual
expansion of the universe at dif-
epochs, as determined by the red
uminosity relation of the most dis-
galaxies, fail to tell us unambigu-
whether the expansion will finally
nd be reversed or whether it will
ue indefinitely.

clumping of matter into stars,
es and clusters of galaxies in the
niverse might seem to make Fried-
s models, based on perfect homo-
y, empty exercises. In actuality the
iness" we observe in the universe
uch a small scale that Friedmann's
ns remain valid. The reason is that
avitational influence of local irreg-
es is swamped by that of more dis-
atter.

haps the most convincing evidence
pport of Friedmann's simple de-
on of the universe was supplied

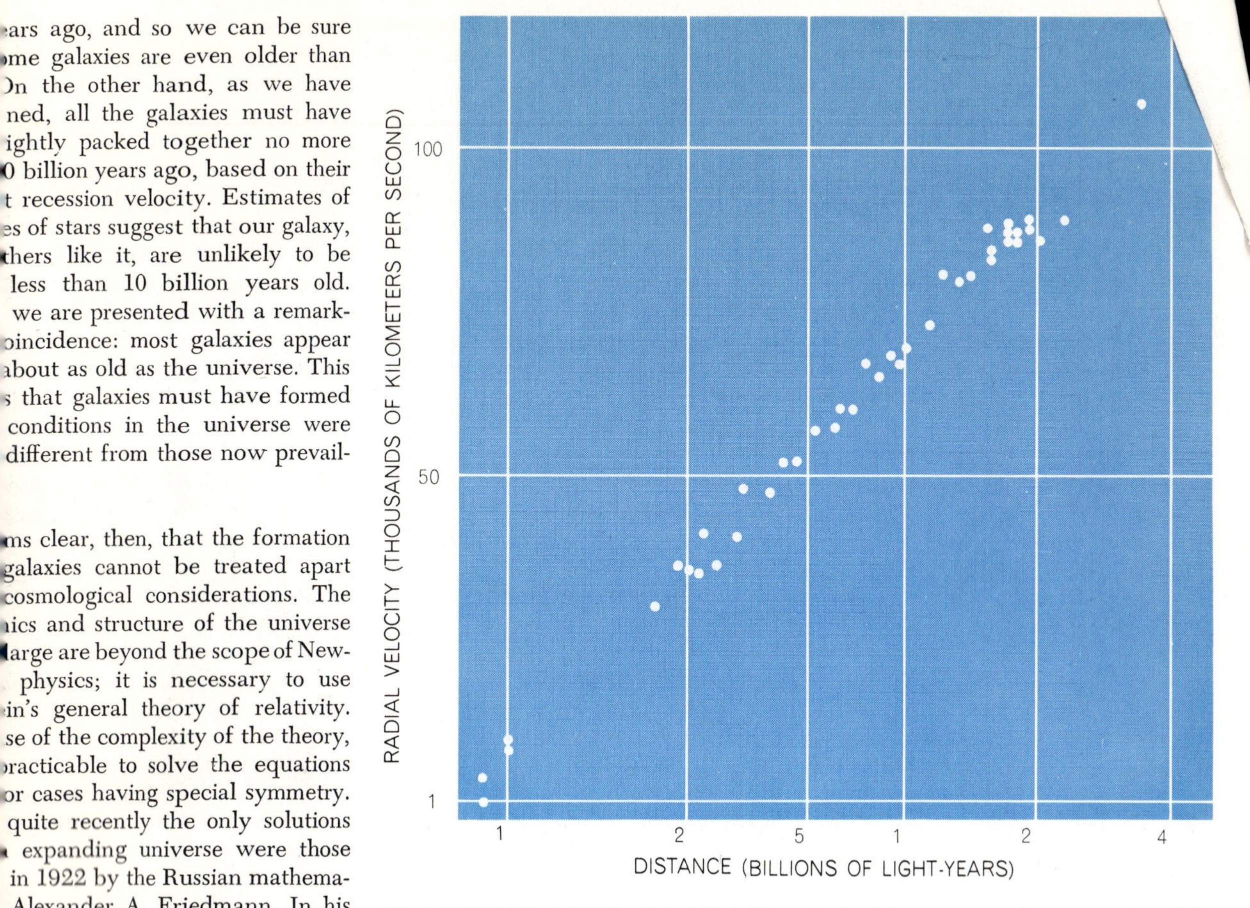

RECESSION VELOCITY OF A GALAXY is obtained by measuring the amount by which the radiation it emits is shifted to the red end of the spectrum. The velocity is directly proportional to the galaxy's distance, as judged by its apparent luminosity. In this diagram, adapted from a recent study by Allan R. Sandage of the Hale Observatories, the ratio of recession velocity to distance is shown for brightest galaxy in each of 41 clusters of galaxies.

in 1965 by the discovery that space is pervaded by a background radiation that peaks at the microwave wavelength of about two millimeters, corresponding to the radiation emitted by a black body at an absolute temperature of three degrees (three degrees Kelvin). This radiation could be the remnant "whisper" from the big bang of creation. The remarkable isotropy, or nondirectionality, of this radiation is impressive evidence for the isotropy of the universe.

The radiation was discovered independently and almost simultaneously at the Bell Telephone Laboratories and at Princeton University [see "The Primeval Fireball," by P. J. E. Peebles and David T. Wilkinson; SCIENTIFIC AMERICAN, June, 1967]. The radiation has the spectrum characteristic of radiation that has attained thermal equilibrium with its surroundings as a result of repeated absorption and reemission, and it is generally interpreted as being a relic of a time when the entire universe was hot, dense and opaque. The radiation would have cooled and shifted toward longer wavelengths in the course of the universal expansion but would have retained a thermal spectrum. It thus constitutes remarkably direct evidence for the hot-big-bang model of the universe first examined in detail by George Gamow in 1940.

Assuming the general validity of the Friedmann model for the early stages of the universe, it seems clear that the material destined to condense into galaxies cannot always have been in discrete lumps but may have existed merely as slight enhancements above the mean density. There will be a tendency for the larger irregularities to be amplified simply because, on sufficiently large scales, gravitational forces predominate over pressure forces that tend to oppose collapse. This phenomenon, known as gravitational instability, was recognized by Newton, who, in a letter to Richard Bentley, the Master of Trinity College, wrote:

"It seems to me, that if the matter of our sun and planets, and all the matter of the Universe, were evenly scattered

The Origin of Galaxies 7

by Martin J. Rees and Joseph Silk
June 1970

The size, shape and other properties of the observed galaxies are traced to slight enhancements in the expanding primordial fireball. Enhancements of certain mass were favored over others

Perhaps the most startling discovery made in astronomy this century is that the universe is populated by billions of galaxies and that they are systematically receding from one another, like raisins in an expanding pudding. If galaxies had always moved with their present velocities, they would have been crowded on top of one another about 10 billion years ago. This simple calculation has led to the cosmological hypothesis that the world began with the explosion of a primordial atom containing all the matter in the universe. A quite different line of speculation argues that the universe has always looked as it does now, that new matter is continuously being created and that new galaxies are formed to replace those that disappear over the "horizon."

On either hypothesis it is still necessary to account for the formation of galaxies. Why does matter tend to aggregate in bundles of this particular size? Why do galaxies comprise a limited hierarchy of shapes? Why do spiral galaxies rotate like giant pinwheels? Astrophysicists are trying to answer these and similar questions from first principles. The goal is to explain as many aspects of the universe as one can without invoking special conditions at the time of origin. In most of what follows we shall assume a cosmological model in which the universe starts with a "big bang." When we have finished, the reader will see, however, that some form of continuous creation of matter may not be ruled out.

Before the invention of the telescope the unaided human eye could see between 5,000 and 10,000 stars, counting all those visible in different seasons. Even modest telescopes revealed millions of stars and in addition disclosed the existence of many diffuse patches of light, not at all like stars. These extragalactic "nebulas," many of them beautiful spirals, are seen in all directions and in great profusion. As early as the 18th century Sir William Herschel and Immanuel Kant suggested that these nebulas were actually "island universes," huge aggregations of stars lying far beyond the limits of the Milky Way.

The validity of this hypothesis was not confirmed until 1924, when the American astronomer Edwin P. Hubble succeeded in measuring the distances to a number of spiral nebulas. Several years earlier Henrietta S. Leavitt had shown that Cepheid variables, named for the prototype Delta Cephei, a variable star discovered in 1784, had light curves that could be correlated with their magnitude. The distances of a number of Cepheids were later determined by independent means, so that it became possible to use more distant Cepheids as "standard candles" to establish a distance-magnitude relation. Hubble looked for Cepheid variables in some of the nearer external galaxies and found them. From their period he was able to deduce their absolute luminosity, and from this he was able to estimate their distance. Hubble soon established that the nearest spiral nebulas (or galaxies) were vast systems of stars situated a million or more light-years outside our own galaxy.

Subsequently Hubble developed a scheme for classifying galaxies according to their morphology, ranging from systems that are amorphous, reddish and elliptical to systems that are highly flattened disks with a complex spiral structure containing many blue stars and lanes of gas and dust [*see illustration on page 55*]. The spiral galaxies themselves vary in appearance. At one extreme are those with large, bright nuclei and inconspicuous, tightly coiled spiral arms. At the other extreme are galaxies in which the nuclei are less dominant and the spiral arms are loosely wound and prominent. The elliptical galaxies also form a sequence, ranging from almost spherical systems to flattened ellipsoids. In addition there are highly irregular systems showing very little structure of any kind.

In all these sequences there is a parallel progression in certain characteristics of the galaxies. In general spirals are rich in gas and dust, contain many blue supergiant stars, are highly flattened and rotate appreciably. Ellipticals, by contrast, seem to possess little gas or dust, usually contain late-type dwarf stars and exhibit scant rotation.

The masses of galaxies are found by several methods. Galaxies are often gravitationally bound together in pairs. If the distance between them and their relative velocities are known, Kepler's law can be used to find their total mass.

⟶

CLUSTER OF GALAXIES in the constellation of Hercules demonstrates the inhomogeneity of the distribution of galaxies in the sky. About 350 million light-years away, this cluster contains about 100 members and is some five million light-years across. It was photographed with the 200-inch Hale telescope on Palomar Mountain. Some very rich clusters contain 1,000 members or more and vary from one million to 10 million light-years across. There is some evidence that such clusters are in turn grouped together into superclusters of perhaps 100 members, spread over 100 million light-years. On scales larger than this the universe appears to be uniform. The bright circular spots with the spikes radiating from them are nearby stars; the spikes are produced by reflections within the telescope.

Another method, used mostly for spirals that are viewed edge on or obliquely, is to determine the velocity of rotation by measuring the Doppler shift of spectral lines emitted by ionized gas in various parts of the disk. (The spectral lines of approaching gas will be shifted toward the blue end of the spectrum, those of retreating gas toward the red end.) One can plot a rotation curve showing how the velocity of rotation varies with the distance from the center of the galaxy. The mass can then be estimated from the requirement that the centrifugal and gravitational (centripetal) forces must be in balance. It turns out that the masses of galaxies are typically about 10^{11} (100 billion) times the mass of the sun. The range, however, is fairly broad: from about 10^8 solar masses for some nearby dwarf galaxies to 10^{12} solar masses for giant ellipticals in more remote regions of the universe. The diameter of the larger spirals, such as our own galaxy, is about 100,000 light-years.

Galaxies also differ widely in the ratio of mass to luminosity. Taking the mass-to-luminosity ratio of the sun as unity, one finds that for large spirals, such as our own galaxy, the ratio varies from one up to 10. In other words, some spirals emit only a tenth as much light per unit of mass as the sun does. Ellipticals commonly emit even less: only about a fiftieth as much light per unit of mass. (Thus their mass-luminosity ratio is 50.)

The distribution of galaxies in the sky is quite inhomogeneous. There are many small groups of galaxies, and here and there some rich clusters containing up to 1,000 members or more. Such systems vary from one million light-years across to 10 million. Our own galaxy is a member of the "local group," an association of about 20 galaxies, only one of which, the Andromeda galaxy, has a mass comparable to that of ours. The local group is about three million light-years in diameter. The Andromeda galaxy is some two million light-years away; the nearest large cluster of galaxies is in Virgo, about 30 million light-years distant.

Even such clusters do not seem to be randomly distributed in space. Some astronomers have argued that there is evidence that clusters are grouped into superclusters of pe
spread over 100 m
universe appears t
larger than this.

Establishing the
was only part
ment. Working w
scope on Mount
from red-shift m
galaxies are in rec
moreover, that th
is directly propor
as judged by its
The most distant
in a faint cluster
Boötes; Rudolph
that the wavelen
from this cluster i
cent. The corresp
cession is nearly h
Light originating
liant starlike obje
is red-shifted mo
but astronomers d
this red shift is du
expansion of the u

The light from
galaxies set out to

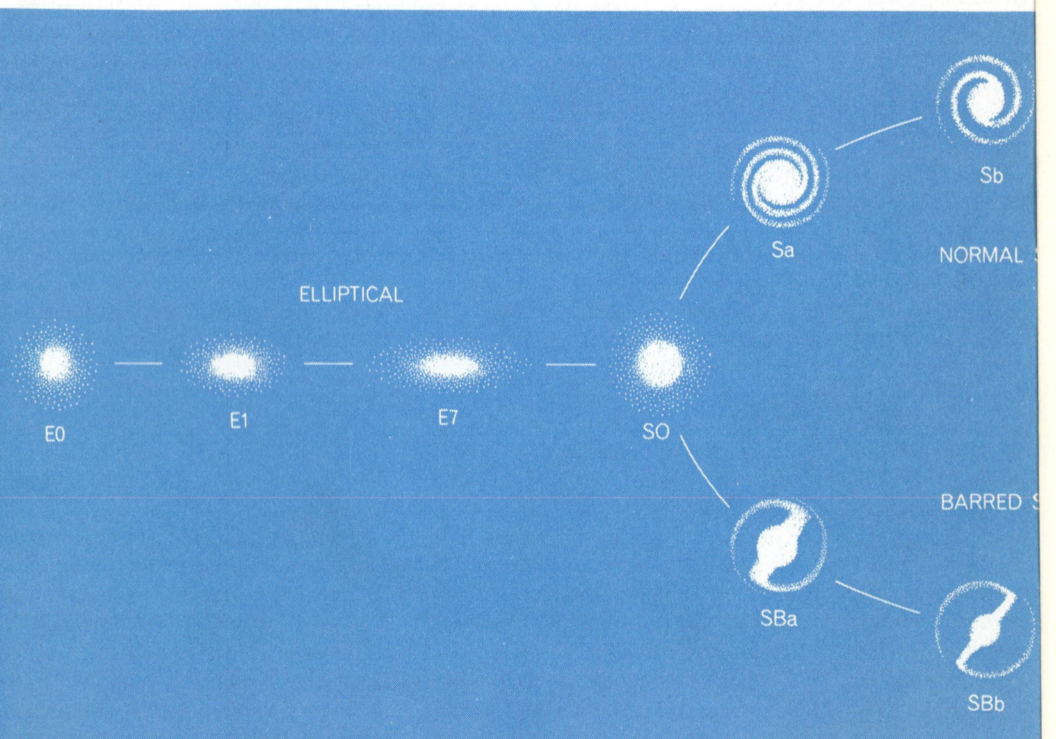

CLASSIFICATION SCHEME developed by Edwin P. Hubble in the early 1930's organizes galaxies according to their shape, ranging from amorphous elliptical systems containing many red stars and little gas and dust (*left*) to highly flattened spiral disks containing many blue stars and lanes of gas and dust (*right*). The elliptical galaxies range from almost spherical systems (designated E0) to highly flattened ellipsoids (E7). The spiral galaxies themselves form two sequences: normal spirals (*top ri*
(*bottom right*). At one extreme in both sequ
large bright nuclei and inconspicuous, tigh
(Sa, SBa); at the other extreme are galaxi
are less dominant and the spiral arms ar
prominent (Sc, SBc). At the branching poi
disklike form that resembles the spirals but

The Origin of Galaxies 7

by Martin J. Rees and Joseph Silk
June 1970

The size, shape and other properties of the observed galaxies are traced to slight enhancements in the expanding primordial fireball. Enhancements of certain mass were favored over others

Perhaps the most startling discovery made in astronomy this century is that the universe is populated by billions of galaxies and that they are systematically receding from one another, like raisins in an expanding pudding. If galaxies had always moved with their present velocities, they would have been crowded on top of one another about 10 billion years ago. This simple calculation has led to the cosmological hypothesis that the world began with the explosion of a primordial atom containing all the matter in the universe. A quite different line of speculation argues that the universe has always looked as it does now, that new matter is continuously being created and that new galaxies are formed to replace those that disappear over the "horizon."

On either hypothesis it is still necessary to account for the formation of galaxies. Why does matter tend to aggregate in bundles of this particular size? Why do galaxies comprise a limited hierarchy of shapes? Why do spiral galaxies rotate like giant pinwheels? Astrophysicists are trying to answer these and similar questions from first principles. The goal is to explain as many aspects of the universe as one can without invoking special conditions at the time of origin. In most of what follows we shall assume a cosmological model in which the universe starts with a "big bang." When we have finished, the reader will see, however, that some form of continuous creation of matter may not be ruled out.

Before the invention of the telescope the unaided human eye could see between 5,000 and 10,000 stars, counting all those visible in different seasons. Even modest telescopes revealed millions of stars and in addition disclosed the existence of many diffuse patches of light, not at all like stars. These extragalactic "nebulas," many of them beautiful spirals, are seen in all directions and in great profusion. As early as the 18th century Sir William Herschel and Immanuel Kant suggested that these nebulas were actually "island universes," huge aggregations of stars lying far beyond the limits of the Milky Way.

The validity of this hypothesis was not confirmed until 1924, when the American astronomer Edwin P. Hubble succeeded in measuring the distances to a number of spiral nebulas. Several years earlier Henrietta S. Leavitt had shown that Cepheid variables, named for the prototype Delta Cephei, a variable star discovered in 1784, had light curves that could be correlated with their magnitude. The distances of a number of Cepheids were later determined by independent means, so that it became possible to use more distant Cepheids as "standard candles" to establish a distance-magnitude relation. Hubble looked for Cepheid variables in some of the nearer external galaxies and found them. From their period he was able to deduce their absolute luminosity, and from this he was able to estimate their distance. Hubble soon established that the nearest spiral nebulas (or galaxies) were vast systems of stars situated a million or more light-years outside our own galaxy.

Subsequently Hubble developed a scheme for classifying galaxies according to their morphology, ranging from systems that are amorphous, reddish and elliptical to systems that are highly flattened disks with a complex spiral structure containing many blue stars and lanes of gas and dust [*see illustration on page 55*]. The spiral galaxies themselves vary in appearance. At one extreme are those with large, bright nuclei and inconspicuous, tightly coiled spiral arms. At the other extreme are galaxies in which the nuclei are less dominant and the spiral arms are loosely wound and prominent. The elliptical galaxies also form a sequence, ranging from almost spherical systems to flattened ellipsoids. In addition there are highly irregular systems showing very little structure of any kind.

In all these sequences there is a parallel progression in certain characteristics of the galaxies. In general spirals are rich in gas and dust, contain many blue supergiant stars, are highly flattened and rotate appreciably. Ellipticals, by contrast, seem to possess little gas or dust, usually contain late-type dwarf stars and exhibit scant rotation.

The masses of galaxies are found by several methods. Galaxies are often gravitationally bound together in pairs. If the distance between them and their relative velocities are known, Kepler's law can be used to find their total mass.

⟶

CLUSTER OF GALAXIES in the constellation of Hercules demonstrates the inhomogeneity of the distribution of galaxies in the sky. About 350 million light-years away, this cluster contains about 100 members and is some five million light-years across. It was photographed with the 200-inch Hale telescope on Palomar Mountain. Some very rich clusters contain 1,000 members or more and vary from one million to 10 million light-years across. There is some evidence that such clusters are in turn grouped together into superclusters of perhaps 100 members, spread over 100 million light-years. On scales larger than this the universe appears to be uniform. The bright circular spots with the spikes radiating from them are nearby stars; the spikes are produced by reflections within the telescope.

Another method, used mostly for spirals that are viewed edge on or obliquely, is to determine the velocity of rotation by measuring the Doppler shift of spectral lines emitted by ionized gas in various parts of the disk. (The spectral lines of approaching gas will be shifted toward the blue end of the spectrum, those of retreating gas toward the red end.) One can plot a rotation curve showing how the velocity of rotation varies with the distance from the center of the galaxy. The mass can then be estimated from the requirement that the centrifugal and gravitational (centripetal) forces must be in balance. It turns out that the masses of galaxies are typically about 10^{11} (100 billion) times the mass of the sun. The range, however, is fairly broad: from about 10^8 solar masses for some nearby dwarf galaxies to 10^{12} solar masses for giant ellipticals in more remote regions of the universe. The diameter of the larger spirals, such as our own galaxy, is about 100,000 light-years.

Galaxies also differ widely in the ratio of mass to luminosity. Taking the mass-to-luminosity ratio of the sun as unity, one finds that for large spirals, such as our own galaxy, the ratio varies from one up to 10. In other words, some spirals emit only a tenth as much light per unit of mass as the sun does. Ellipticals commonly emit even less: only about a fiftieth as much light per unit of mass. (Thus their mass-luminosity ratio is 50.)

The distribution of galaxies in the sky is quite inhomogeneous. There are many small groups of galaxies, and here and there some rich clusters containing up to 1,000 members or more. Such systems vary from one million light-years across to 10 million. Our own galaxy is a member of the "local group," an association of about 20 galaxies, only one of which, the Andromeda galaxy, has a mass comparable to that of ours. The local group is about three million light-years in diameter. The Andromeda galaxy is some two million light-years away; the nearest large cluster of galaxies is in Virgo, about 30 million light-years distant.

Even such clusters do not seem to be randomly distributed in space. Some astronomers have argued that there is evidence that clusters are grouped into superclusters of perhaps 100 members, spread over 100 million light-years. The universe appears to be uniform on scales larger than this.

Establishing the distance of galaxies was only part of Hubble's achievement. Working with the 100-inch telescope on Mount Wilson, he showed from red-shift measurements that the galaxies are in recession. Hubble found, moreover, that the red shift of a galaxy is directly proportional to its distance, as judged by its apparent luminosity. The most distant galaxies known are in a faint cluster in the constellation Boötes; Rudolph Minkowski discovered that the wavelength of light coming from this cluster is stretched by 45 percent. The corresponding velocity of recession is nearly half the speed of light. Light originating from some of the brilliant starlike objects known as quasars is red-shifted more than 200 percent, but astronomers disagree whether or not this red shift is due to the cosmological expansion of the universe.

The light from Minkowski's cluster of galaxies set out toward us about five bil-

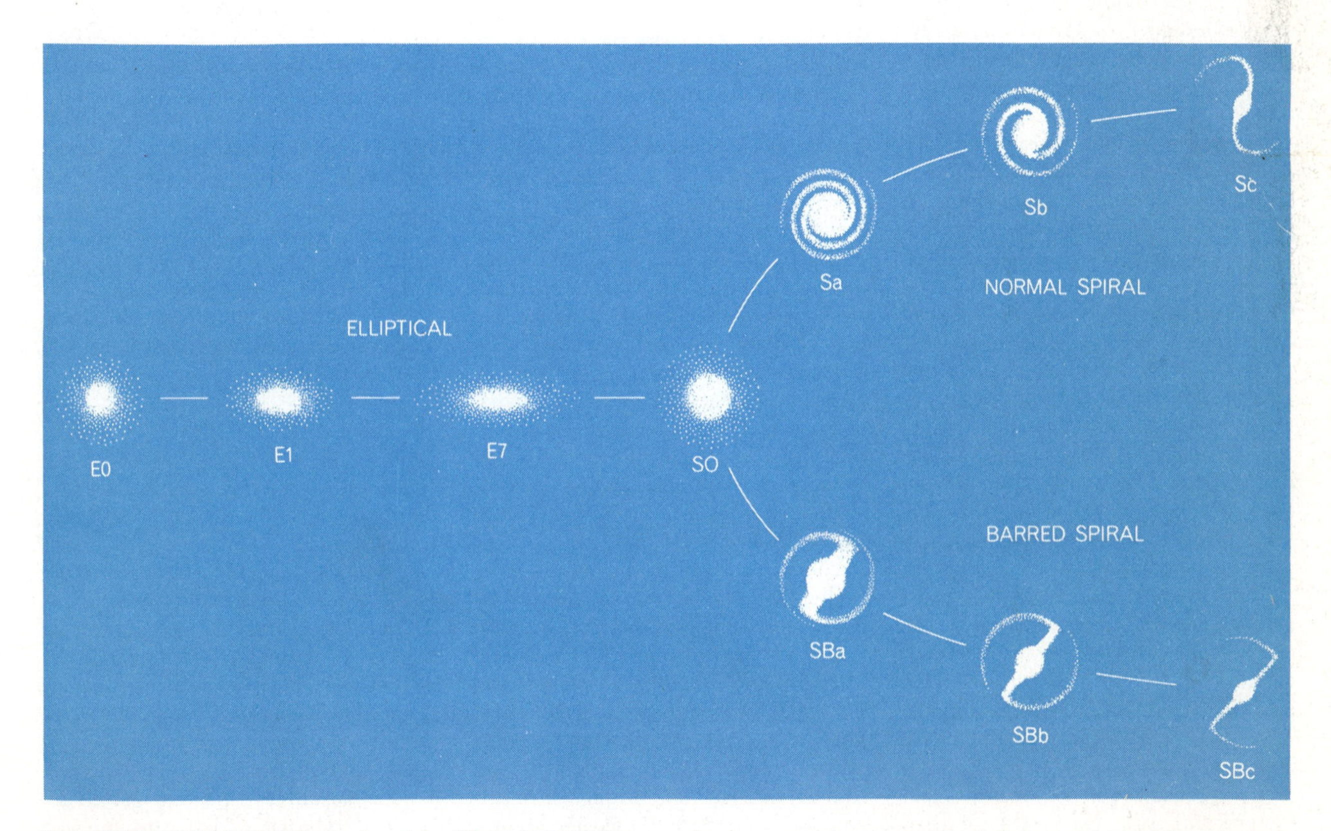

CLASSIFICATION SCHEME developed by Edwin P. Hubble in the early 1930's organizes galaxies according to their shape, ranging from amorphous elliptical systems containing many red stars and little gas and dust (*left*) to highly flattened spiral disks containing many blue stars and lanes of gas and dust (*right*). The elliptical galaxies range from almost spherical systems (designated E0) to highly flattened ellipsoids (E7). The spiral galaxies themselves form two sequences: normal spirals (*top right*) and barred spirals (*bottom right*). At one extreme in both sequences are galaxies with large bright nuclei and inconspicuous, tightly coiled spiral arms (Sa, SBa); at the other extreme are galaxies in which the nuclei are less dominant and the spiral arms are loosely wound and prominent (Sc, SBc). At the branching point of the diagram is a disklike form that resembles the spirals but lacks spiral arms (SO).

lion years ago, and so we can be sure that some galaxies are even older than that. On the other hand, as we have mentioned, all the galaxies must have been tightly packed together no more than 10 billion years ago, based on their present recession velocity. Estimates of the ages of stars suggest that our galaxy, and others like it, are unlikely to be much less than 10 billion years old. Hence we are presented with a remarkable coincidence: most galaxies appear to be about as old as the universe. This implies that galaxies must have formed when conditions in the universe were much different from those now prevailing.

It seems clear, then, that the formation of galaxies cannot be treated apart from cosmological considerations. The dynamics and structure of the universe in the large are beyond the scope of Newtonian physics; it is necessary to use Einstein's general theory of relativity. Because of the complexity of the theory, it is practicable to solve the equations only for cases having special symmetry. Until quite recently the only solutions for an expanding universe were those found in 1922 by the Russian mathematician Alexander A. Friedmann. In his idealized models matter is treated as a strictly uniform and homogeneous medium. The universe expands from a singular state of infinite density, with the rate of expansion decelerating as a consequence of the mutual gravitational attraction of its different parts. The universe may have enough energy to keep expanding indefinitely or the expansion may eventually cease and be followed by a general collapse back to a compressed state. Observations of the actual rate of expansion of the universe at different epochs, as determined by the red shift–luminosity relation of the most distant galaxies, fail to tell us unambiguously whether the expansion will finally stop and be reversed or whether it will continue indefinitely.

The clumping of matter into stars, galaxies and clusters of galaxies in the real universe might seem to make Friedmann's models, based on perfect homogeneity, empty exercises. In actuality the "graininess" we observe in the universe is on such a small scale that Friedmann's solutions remain valid. The reason is that the gravitational influence of local irregularities is swamped by that of more distant matter.

Perhaps the most convincing evidence in support of Friedmann's simple description of the universe was supplied in 1965 by the discovery that space is pervaded by a background radiation that peaks at the microwave wavelength of about two millimeters, corresponding to the radiation emitted by a black body at an absolute temperature of three degrees (three degrees Kelvin). This radiation could be the remnant "whisper" from the big bang of creation. The remarkable isotropy, or nondirectionality, of this radiation is impressive evidence for the isotropy of the universe.

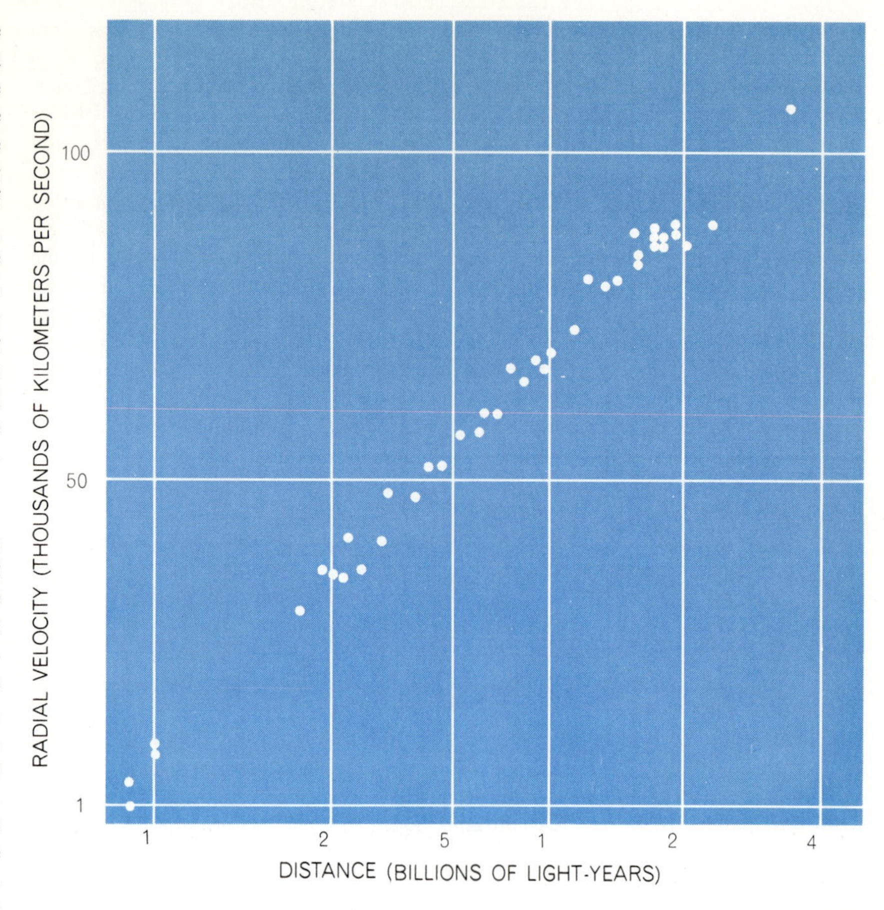

RECESSION VELOCITY OF A GALAXY is obtained by measuring the amount by which the radiation it emits is shifted to the red end of the spectrum. The velocity is directly proportional to the galaxy's distance, as judged by its apparent luminosity. In this diagram, adapted from a recent study by Allan R. Sandage of the Hale Observatories, the ratio of recession velocity to distance is shown for brightest galaxy in each of 41 clusters of galaxies.

The radiation was discovered independently and almost simultaneously at the Bell Telephone Laboratories and at Princeton University [see "The Primeval Fireball," by P. J. E. Peebles and David T. Wilkinson; SCIENTIFIC AMERICAN, June, 1967]. The radiation has the spectrum characteristic of radiation that has attained thermal equilibrium with its surroundings as a result of repeated absorption and reemission, and it is generally interpreted as being a relic of a time when the entire universe was hot, dense and opaque. The radiation would have cooled and shifted toward longer wavelengths in the course of the universal expansion but would have retained a thermal spectrum. It thus constitutes remarkably direct evidence for the hot-big-bang model of the universe first examined in detail by George Gamow in 1940.

Assuming the general validity of the Friedmann model for the early stages of the universe, it seems clear that the material destined to condense into galaxies cannot always have been in discrete lumps but may have existed merely as slight enhancements above the mean density. There will be a tendency for the larger irregularities to be amplified simply because, on sufficiently large scales, gravitational forces predominate over pressure forces that tend to oppose collapse. This phenomenon, known as gravitational instability, was recognized by Newton, who, in a letter to Richard Bentley, the Master of Trinity College, wrote:

"It seems to me, that if the matter of our sun and planets, and all the matter of the Universe, were evenly scattered